Crowds in Equations

An Introduction to the Microscopic Modeling of Crowds

Advanced Textbooks in Mathematics

Print ISSN: 2059-769X
Online ISSN: 2059-7703

The *Advanced Textbooks in Mathematics* explores important topics for post-graduate students in pure and applied mathematics. Subjects covered within this textbook series cover key fields which appear on MSc, MRes, PhD and other multidisciplinary postgraduate courses which involve mathematics.

Written by senior academics and lecturers recognised for their teaching skills, these textbooks offer a precise, introductory approach to advanced mathematical theories and concepts, including probability theory, statistics and computational methods.

Published

Crowds in Equations: An Introduction to the Microscopic Modeling of Crowds
   by Bertrand Maury and Sylvain Faure

The Wigner Transform
   by Maurice de Gosson

Periods and Special Functions in Transcendence
   by Paula B Tretkoff

Mathematics of Planet Earth: A Primer
   by Jochen Bröcker, Ben Calderhead, Davoud Cheraghi, Colin Cotter,
   Darryl Holm, Tobias Kuna, Beatrice Pelloni, Ted Shepherd and Hilary Weller
   edited by Dan Crisan

Forthcoming

Conformal Maps and Geometry
   by Dmitry Beliaev

Advanced Textbooks in Mathematics

Crowds in Equations

An Introduction to the Microscopic Modeling of Crowds

Bertrand Maury

Université Paris-Sud, France

Sylvain Faure

Université Paris-Sud & CNRS, France

World Scientific

NEW JERSEY · LONDON · SINGAPORE · BEIJING · SHANGHAI · HONG KONG · TAIPEI · CHENNAI · TOKYO

Published by

World Scientific Publishing Europe Ltd.

57 Shelton Street, Covent Garden, London WC2H 9HE

Head office: 5 Toh Tuck Link, Singapore 596224

USA office: 27 Warren Street, Suite 401-402, Hackensack, NJ 07601

Library of Congress Cataloging-in-Publication Data

Names: Maury, Bertrand, author. | Faure, Sylvain, 1976–   author.

Title: Crowds in equations : an introduction to the microscopic modeling of crowds / by
    Bertrand Maury (Université Paris-Sud, France),
    Sylvain Faure (Université Paris-Sud & CNRS, France).

Description: New Jersey : World Scientific, 2018. | Series: Advanced textbooks in mathematics |
    Includes bibliographical references and index.

Identifiers: LCCN 2018013165 | ISBN 9781786345516 (hc : alk. paper)

Subjects: LCSH: Mathematical analysis. | Mathematical models. | Communication in mathematics.

Classification: LCC QA401 .M448 2018 | DDC 511/.8--dc23

LC record available at https://lccn.loc.gov/2018013165

British Library Cataloguing-in-Publication Data

A catalogue record for this book is available from the British Library.

For any available supplementary material, please visit
http://www.worldscientific.com/worldscibooks/10.1142/Q0163#t=suppl

Desk Editors: V. Vishnu Mohan/Jennifer Brough/Koe Shi Ying

Typeset by Stallion Press
Email: enquiries@stallionpress.com

Foreword

Building bridges, remarking similarities, crossing methods are essential driving forces of the scientific activity. Since ancient times, physics and mathematics have been interwoven. Today, although the amount of knowledge makes it essentially impossible to have a global view on these so-called hard sciences, interactions remain fruitful. On the other hand, social sciences are — at first sight — unconcerned with this thinking system: they are interested in a multitude of behaviours which seem to escape the deterministic laws of physics and any mathematical formulation.

A small revolution of the last decade has been to remove this barrier to allow a new research field to emerge. If one observes typical behaviours (at the scale of many individuals) and that one can identify parameters influencing these behaviours, then there is no reason that the mathematical language cannot describe them! Of course, the connection is not so easy and this new field is still in its infancy, not always considered seriously by its elders.

In this book, intended to graduate students and researchers in mathematics, Sylvain Faure and Bertrand Maury invite us to discover the challenges and the first successes of mathematics applied to social sciences. As a preamble, they clearly explain the difficulties of the exercise, due, in particular, to the freedom of individuals and to the decision processes which are neither symmetric, nor interchangeable.

The book continues with the rigorous analysis of some models, essentially at the microscopic scale, which serve as mathematical prototypes and exhibit interesting phenomenologies.

Starting from this picture, the authors propose eventually to extract the minimal elements which should be contained in a mathematical model in order to reproduce some typical and sometimes paradoxical properties of crowd motions: "Faster-is-Slower" effect, "Stop-and-Go" waves, and fluidizing effects of an obstacle. This is fascinating! A quick and easy read, which makes me want to learn more.

Laure Saint-Raymond

École Normale Supérieure de Lyon &
Académie des Sciences, France

Contents

Chapter 1

Introduction

1.1. From Passive to Active Entities

The modeling of particle systems has raised a considerable activity in the last centuries. Physicists, mathematicians, and more recently computer scientists, have joined their efforts to formalize the laws which govern the motion of particles in interaction with one another. They built theoretical frameworks and numerical tools to evolve the "big picture", i.e. induce general rules or macroscopic equations which would make it possible to describe the behavior of the considered system at a large scale, beyond the individual destinies of its components.

Since a few decades, scientists from various domains have extended the approach to systems of "active entities", bird flocks, fish schools, crowds of mammals or insects, and, even further apart from passive particles, walking or driving human beings. Part of the frameworks developed for physical systems can be straightforwardly transposed to this new situation. In particular, the kinematic modeling will consist in representing a moving crowd by a time-varying vector which contains the positions of its individuals. Now assuming one is able to model the will of an individual and its interaction with others in an equation involving the positions and its derivatives, the model takes the form of a system of differential equations, for which a tremendous amount of theoretical and numerical tools have been developed. Yet, the modeling of living entities presents particular features which make it very peculiar among other activities in particle system modeling.

(1) (**Out of equilibrium thermodynamics**) The fact that the considered entities are active from the dynamical point of view (they are capable of using their internal energy for their own motion) rules out the possibility to expect some kind of thermalization of the overall system.

(2) (**Decision processes**) Human beings are also *active* in a decisional sense. Whereas passive particles obey mechanical laws, and preprogrammed automata evolve according to predetermined rules, human beings have the ability to design their own instantaneous strategies. Such choices may be based on a full or partial knowledge of their environment (e.g. positions and possibly velocities of neighbors), and also triggered by emotional factors (impatience, stress, justified or artificial panic, . . .).

(3) (**Asymmetric interactions**) As mentioned in the previous point, human agents take decisions according to their instantaneous knowledge. In the context of crowd motion, the latter mostly come from *vision*. Since vision is oriented (pedestrians typically visualize a certain cone about the direction they head to), the influence relations within a crowd are also oriented, which kills the symmetry of interactions. Although the notion of *social force* is commonly used to account for interactions, it must not be forgotten that those so-called forces may rule out the Law of Action–Reaction which governs the dynamics of passive particles.

(4) (**Mesoscopic scale**) Some important features in crowd motion develop at a mesoscopic scale, i.e. they appear and must be described at a scale that is not infinitely large with respect to the microscopic scale (i.e. the size of an individual). As we shall see, it concerns for example "Stop-and-Go" waves in a queue, the wavelength of which merely extends to a few individuals (to be compared to audible sound waves in a gas, the typical length of which scales like 10^8 times the inter-particle distance). In a different context, the evacuation of a room strongly depends on what happens in the very neighborhood of the exit door, the size of which scales like people diameter.

(5) (**Non-interchangeable entities**) Although most models presented in this book will be based on the simplifying assumption that individuals behave in a similar way, it should not be forgotten that real crowds are made of individuals with different personalities (in terms of aggressivity, politeness, ability to develop strategies, tendency to cooperate, . . .).

As such, the state of the system cannot be defined up to permutations, like in the case of gas particles.

1.2. Basics on Crowd Motion Modeling

We shall essentially restrict ourselves in this book to microscopic models. Those are based on an individual tracking of individuals, and they are by nature *Lagrangian*: each variable will be attached to a given individual. Most microscopic models rely on two main ingredients:

(1) Defining individual tendencies, i.e. choosing tractable rules to determine, at each instant, what an individual would do if they were alone. Note that such an issue may not be relevant in some situations, e.g. for a child following his parents, but we shall generally consider that the atomic entity of the model is a responsible person with an own intention. In the case of a building evacuation, which will be used as an archetypal example in this book, such an intention is clearly defined: reach the closest exit in a short time.
(2) Interaction rules: how is a pedestrian influenced by other pedestrians in their neighborhood? The core of each model relies in accounting for those interactions, which may be of various types. In this book we shall focus on two types of interaction: physical interaction between two people in contact, or social tendency to preserve a certain distance to neighbors. The first type of interaction is purely mechanical, and fits in the classical framework of granular mechanics (the term *grains* is preferred to that of *particle* when finite-size effects are significant). The second type is characteristic of living entities, since it is not triggered by a physical contact, but rather implements a complex cognitive process which is initiated by the perception (typically vision) of neighbors. As such, it may disobey the Law of Action–Reaction.

The way those ingredients are encoded in the various models depends on the type of *representation* which is chosen, as detailed in the next two sections.

Individual tendencies

As for individual tendencies, the core notion will be that of *desired velocity*, all over this book. Let us make it clear that this notion is somewhat ambiguous, since one may consider that the instantaneous desire of a civilized pedestrian accounts for people in their neighborhood. Our choice, following that of most authors, will be to favor the selfish meaning of "desired": the desired velocity, which we may also call spontaneous velocity, is the velocity that a person would have *if they were alone*. It therefore corresponds to an instantaneous tendency of achieving a given goal (e.g. exiting a building in fire) in a purely selfish way. In the one-dimensional setting (see Chapter 2), pedestrians are assumed to walk one behind another, all heading to the same direction. Under those assumptions, the desired velocity turns out to be a desired *speed*, or *free speed*. In the two-dimensional setting, the desired direction must also be prescribed. Chapter 8 is dedicated to the question of determining the desired velocity field in various contexts. We shall favor scenarios of the evacuation type, which makes it possible to properly set the problem without too much psychological considerations: the desired velocity of agents will be built as best suited to achieve the common goal, that is to exit the building as fast as possible. In general, this velocity will be defined (up to a multiplicative constant) as the opposite of the gradient of a quantity that a pedestrian tends to minimize. In general, this quantity has a vocation to represent a pedestrian's dissatisfaction, like the distance to the exit for an evacuation.

The actual encoding of the individual tendency depends on the model representation. In the inertial social force model (Chapter 3), it is encoded as a relaxation term: the pedestrian behaves like an inertial particle in a viscous fluid, the velocity of which is the desired velocity. Viscous frictional forces tend to have the pedestrian/particle move at the same velocity than the underlying fluid. The non-inertial version of this model (Section 3.2 in the same chapter) corresponds to the case of an infinitely viscous fluid: the particle behaves like a passive tracker which follows the motion of the underlying fluid. The differential system is then set on the positions only, and the desired velocity is the main forcing term which conditions the actual velocity (corrections due to interactions will be addressed below). In the granular approach (Chapter 4), the set of desired velocity field is considered as some sort of *attempted* velocity field, which is projected to the set of admissible velocities (i.e. velocities that respect the non-overlapping constraint). In the context of Cellular Automata (Chapter 5), pedestrians are considered as tokens occupying the cells of a cartesian grid, and the evolution process is of a stochastic nature. In this context, individual tendencies are implemented as bias to the hopping probabilities toward neighboring cell. In some way, CA implement in a stochastic way the assumption that the desired velocity heads along the opposite of the gradient of a quantity that is to be minimized. In the context of CA, this quantity is called the *floor field*, but it plays the exact same role as the dissatisfaction already mentioned.

Interaction rules

It would be quite presumptuous to suggest that all imaginable social interactions can be encoded in equations. We shall therefore restrict ourselves in this book to simple types of interactions, in particular the social tendency to maintain a certain distance from neighbors, together with the physical non-overlapping rule (two individuals may not occupy the same space at the same time).

In the one-dimensional setting (Chapter 2), with pedestrians walking one behind another, it can be considered that walkers tend to preserve a certain distance to the walker behind in order to avoid a collision in the case where the latter suddenly stops. Expressed in a reciprocal way, this tendency will be encoded in assuming that the speed is an increasing function of the distance to the next walker.

In the general, two-dimensional setting, even in the static or quasi-static situation, the tendency to maintain a certain distance from neighbors will be encoded in various ways, depending on the representation. In the so-called social force model (and its extensions), presented in Chapter 3, this tendency is modeled by a repulsive force, which possibly rules out the Law of Action–Reaction (if cone of visions are accounted for).

Another approach consists in representing individuals by rigid disks (see Chapter 4), and to restrict interactions to actual physical contacts. In this framework, interaction forces appear as mathematical auxiliary variables associated to non-overlapping constraints, namely Lagrange multipliers.

In the cellular automata approach, the representation is different (as already mentioned, an individual is identified to a token placed in a cell of a cartesian grid which covers the zone of interest), but interactions are also treated in a hard way, by simply requiring that two agents are not authorized to share a common cell.

1.3. The Mathematical Standpoint

Since most publications on microscopic crowd models come from the community of physicists or computer scientists, while this book tends to adopt a mathematical standpoint, it is mandatory to make some remarks on the "cultural" differences between those approaches. For a mathematician, a model (in the present context of crowd motion) consists in an equation or a set of equations of two main types: Ordinary Differential Equations (ODEs) involve a finite (but possibly large) number of variables depending on time (seen as a one-dimensional continuum), whereas Partial Differential Equations (PDEs) involve unknown fields defined over a certain zone in the Euclidean space. Such a set of equations will be considered as a sound model (from the mathematical standpoint) if it presents a predictive character, i.e. if one can establish that, given initial conditions (and possibly boundary conditions for PDE's), existence and uniqueness of a solution can be rigorously proven. The term *predictive*, at this stage, does not mean that the solution properly describes in a predictive way any real-life phenomenon, it simply means that, under some assumptions and initial conditions, the model can "produce" something that is properly defined. Establishing this well-posedness usually does not give information on the solution itself (although some existence proofs are *constructive*, and can be used to effectively approximate the solution). Except for very particular situations, the solutions to those equations do not admit analytical expressions, so that numerical computations are necessary. Those computations are based on an initial *discretization* process, necessary to replace infinite-dimensional problems by problems involving a finite number of unknowns, which makes an actual computation tractable. For ODEs, it simply consists in replacing the continuous time line by a discrete set of so-called *time steps*. For PDEs, the space itself is discretized, i.e. the domain is decomposed onto a finite number of small cells, to which unknowns are associated. Parameters pertaining to this discretization process have no meaning in terms of modeling. In this framework, when a computation is presented, it is meant (sometimes implicitly) that the computed solutions that are presented are close to the exact solution. By "close" we mean that it presents the same features, up to being indistinguishable in the eye distance. In particular, it requires that reducing for instance the time step should not significantly affect the computed solution. Numerical Analysis is the branch of applied mathematics dedicated to properly quantify this closeness. Convergence results typically ensure that, when the discretization parameters go to 0, the computed solution converges to the exact one for some appropriate norm. Numerical Analysis is not central in the present book, but all numerical methods which we use are of course covered by such convergence results.

The standpoint of physicists may differ: the modeling phase (establishing the equations) and the discretization phase are not always strictly separated. In particular, it is common practice to give a physical interpretation to what a mathematician would call a discretization parameter. This difference is not a minor one: it reflects a deep difference in what is considered as a Model. We shall present in this book (see e.g. Section 3.1) some ingredients added to ODE models, in such a way that the induced effect strongly depends on the time step, in particular it vanishes when the time step goes to zero. It leads to consider this time step as a modeling parameter, rather than a pure discretization parameter. In this context, the model is no longer a system of ODE's, but an essentially discrete procedure, with a time step that is not intended to go to 0. This philosophy can lead to models which are essentially discrete, like cellular automata (CA, see Chapter 5). Those models consist of a succession of discrete events, namely hopping of particles from one cell to another, in a cartesian grid. This grid is similar to discretization meshes which are used to numerically solve PDE's, yet they are different in essence, because the space step is not intended to be taken smaller and smaller. In the context of CA, it represents the zone typically occupied by a pedestrian. In this context, although the algorithm itself may look similar to some numerical methods used to solve PDE, it is not related to any continuous equation.

Both approaches are of course complementary, and it would not make any sense to rank them in terms of relevance, but it is of the utmost importance to keep in mind the distinction to appreciate with the right criteria the various approaches proposed in the literature.

Do mathematics really matter?

The previous considerations raise a natural question: what is the contribution of mathematics in the modeling process, do they really help better understand the real-life phenomena, the underlying mechanisms which make things happen as they do? This question is not purely rhetorical. First, huge contributions in science have been made without the use of deep or sophisticated mathematics (excluding the basic language that is shared by the so-called *hard sciences*). Second, it cannot be denied that some mathematical developments allegedly issued from modeling questions, as interesting as they may be from the strict mathematical standpoint, do not bring much light on the underlying phenomena. Let us add here (this point is a common source of misunderstanding between communities) that the rigor of the theoretical analysis and the numerical analysis, which is usually a full-time job for applied mathematicians, does not prove anything on the quality of the model in terms of representing the underlying reality.

We nevertheless would like here to advocate for the role of some mathematical developments to better understand and describe physical phenomena, up to complex living systems like crowds.

Existence and uniqueness

Let us start with the pet theme in mathematics, that is proving existence and uniqueness of a solution to a given problem. Let us first say that uniqueness of a solution indicates that no equation is missing, given the considered unknown. More precisely, it tells the modeler that no equation or constraint should be added, and even that none *can* be added (addition of extra equations or constraints would overdetermine the problem, and in general rule out existence). Existence of a solution is, paradoxically, more delicate to interpret. Together with uniqueness, it brings satisfaction to the scientist in rigorously allowing them to talk about a *solution* as something that is properly defined (in a well-defined sense). But the richer contribution to mathematical well-posedness analysis to the understanding a model lies often in darker zones, i.e. when things do not work as they are expected to. The fact that well-posedness cannot be established, or only partially (e.g. the solution of an evolution problem exists for a short time period, but global existence cannot be proven), commonly reflects a defect in modeling assumptions, or emergence of a critical situation that is not covered by the model. Let us give an example: in the context of microscopic crowd models, individuals are commonly represented by disks, and interactions between two individuals are encoded by vectors "acting" on both along the line joining their centers. Obviously, the direction of this line smoothly depends on the center location when both are well separated, but the smoothness is lost when those centers get closer up to coincide. Such a proximity up to contact will rule out the possibility to use Cauchy–Lipschitz theory to ensure existence and uniqueness of a global solution. If a model is such that this coincidence is likely to happen, it also means that it authorizes pedestrian to go across each other, so that the theoretical defect actually reflects a flaw in the model itself. Who is not aware of this problem (which is typically due to a bad calibration of repulsion forces, which are supposed to prevent full overlapping from happening) takes the risk to obtain unrealistic results by a straight computation of this ill-posed model. Another defect in well-posedness theory may be due to non-regular desired velocity fields. We shall describe, in the context of building evacuation, how such fields can be built in a natural way, by simply stating that the local velocity corresponds to the shortest path to the closest exit (expressed in a fancy way, the velocity shall be defined as the *opposite of the gradient of the geodesic distance to the set of exits*). A velocity field built in such a way may be not regular as soon as the room contains obstacles, or if there is more than one exit. This will again rule out a straight use of standard theorems to prove existence and uniqueness of a solution. Again, this theoretical defect reflects a feature of the model. The zones where the field is not regular correspond to points from which there is two or more ways to exit the building. It obviously calls for an enrichment of the model in the neighborhood of those singular zone, e.g. by stating that an individual located in this neighborhood shall randomly pick one of the possible ways to exit the building, or will pick the one they took to enter the building.

Stability analysis

Mathematical developments may provide more direct contributions to understanding emergence of some phenomena in the neighborhood of some particular states, and/or to assess the dependence of some observable quantities with respect to model parameters. As an illustration, crowd motion models in a one-dimensional setting (people walking behind each other) are described in Chapter 2. They exhibit simple stationary states, in the neighborhood of which a full stability analysis can be carried out. This analysis consists in investigating the spectrum (set of eigenvalues) of the matrix associated to the linearized system at the equilibrium point. In the complex plane, the zone occupied by this spectrum with respect to the imaginary axis determines the stability of the equilibrium point. For models which are first order in time, the fact that the spectrum lies on the left-hand side of this axis ensures stability. The reciprocal of the real parts of the eigenvalues corresponds to characteristic times, in particular the largest of those times gives the order of magnitude of the time that is taken by a perturbation of the equilibrium state to damp back to equilibrium. The corresponding imaginary parts can be interpreted as time frequencies of damped oscillations undergone by individuals after the system has been perturbed.

More strikingly, the same analysis applied to the inertial (or delayed) version of this model will exhibit native instabilities, i.e. spontaneous emergence of perturbations which tend to propagate upstream (Stop-and-Go waves).

Identifying mechanisms, abstract causes

The present book contains various models of crowd motion, mostly issued from microscopic considerations: those models are built by expressing the behavior of a pedestrian, made of individual tendencies together with ingredients to mimic interactions with other pedestrians. On the other hand, some collective phenomena have been experimentally observed, some of which are paradoxical, like the Capacity Drop (CD) phenomenon, the Faster-is-Slower (FiS) effect, and the fluidizing role of an obstacle. Determining whether or not a given model is capable[1] of reproducing those phenomena have given rise to an important literature (see Chapter 9). Yet, the issue regarding which particular ingredients in the model make it able to reproduce a phenomenon is rarely addressed.

Mathematics can help better understand the very role played by various ingredients, thereby shedding some light on the possible causes of the observed phenomena. In this spirit, a full chapter (Chapter 10) is dedicated to this approach, which we may designate as "abstract inverse modeling", since it consists in determining which ingredients are required for a general model to reproduce a given phenomenon. This is done in Chapter 10 by designing some sorts of minimal models, i.e. equations based on few variables, or minimal sets of equations, which make it possible to recover the phenomenon. The approach is inductive, and does not commonly lead to a single answer. As an illustration, let us mention the Faster-is-Slower effect, which essentially states that, when people evacuate through a narrow exit, it may happen that increasing the velocity (or eagerness) of some of them may harm the overall evacuation process. The approach which we propose makes it possible to single out two main "ingredients" likely to explain the paradoxical effect, each of which alone being able to explain it, that is: the non-convexity of the set of feasible configurations, and friction between individuals.

Modeling = Betraying?

Most models and mathematical developments presented in this book are based on a highly idealized representation of the reality. This feature is

---

[1] Let us make it clear that, since the aforementioned phenomenon are not systematically observed, the non-reproduction of one of those effects does not disqualify a model. We refer the reader to Chapter 9 for a detailed discussion on this matter.

inherent to the very core of mathematical modeling, which favors simplicity and sparsity in terms of parameters. The underlying principle is the following: always elaborate the simplest model which reproduces a given feature. This principle helps in avoiding the curse of overfitting.[2] Besides, the soberness of a model allows for direct use of abstract mathematical results. Yet, this oversimplification has to be questioned:

(1) Do oversimplifying assumptions disqualify the model in its ability to predict the reality in some way, to reproduce characteristic features, and to recover observable quantities with a reasonable accuracy?

(2) Some observable phenomena are explained by abstract mathematical developments. Those mathematical developments may necessitate the model to be "clean and spare". Since the reality is not, does this approach really make sense in terms of modeling?

Those issues go far beyond the scope of this book, but they should be kept in mind to maintain a certain critical thinking while developing new models. We shall restrict ourselves here to shortly instantiate those general questions in connection with crowd motions. To illustrate the first point, consider the representation of pedestrian as two-dimensional simple geometrical objects. Most models indeed rely on the identification of pedestrians to disks, which obviously departs from reality. Some of those models compare sufficiently well with reality to convince that this simplification is not too harmful. Yet, if such a model is meant to be used in a predictive manner, in a situation in which the model has not been validated, there is no guarantee on its ability to properly fulfill its task. As for the second point, let us point to the spectral analysis carried out in Chapter 2 in order to explain the upstream propagation of a perturbation along a line of pedestrians. This analysis relies on strong assumptions in terms of evenness of the pedestrians. Those assumptions are ruled out as soon as one considers that pedestrians may slightly differ in terms of behavior, so that the overall theory (based on the fact that a certain matrix is *not diagonalizable*) becomes inapplicable. We investigate (see p. 27 and following) the possibility to extend the approach dedicated to the homogeneous model to the more realistic case of non-homogeneous populations. This illustrates the necessity to complement the theoretical study of over-simplified models

---

[2]A model with too many parameters, as disrespectful as it may be of the underlying reality, is likely to reproduce any target property.

by an investigation of the *robustness* of the considered models. This investigation aims at comforting the overall modeling approach by checking that the clean and spare model really carries the essential features of the general situation.

1.4.   How to Use this Book?

The various chapters of this book have been conceived as self-contained, and the book is not built in a progressive way. To be more precise, Chapters 2–6, present various independent approaches, all based on a microscopic description of crowds. Each of those chapters can be read at first.

Chapter 7 differs from the previous ones, since it proposes an introduction to macroscopic models, it makes sense to read it after having a fair knowledge in microscopic aspects. It may also be skipped by readers decided to stick to microscopic approaches.

Chapter 8 is more technical and transverse: it is focused on computational aspects. It is in particular dedicated to actual computation of desired velocity fields in non-trivial situations. Since this technique is needed by all approaches, understanding it is mandatory for readers who might want to apply any of the proposed models to real-life situations (for instance in complex buildings).

Chapter 9 is the less mathematically oriented: it is entirely dedicated to experimental evidence, measurable data, and observed phenomena.

Chapter 10 is the most exotic one, it contains free mathematical developments to investigate the possible causes (in the abstract sense of *model features*) of some observed phenomena in crowd motions.

We strongly encourage readers to complement the theoretical study of models by performing their own computations. Numerical algorithms to numerically solve most models in this book are proposed in the Python package `cromosim`, developed by the authors.[3]

---

[3]Python programs used in this book can be obtained at `http://www.cromosim.fr`.

Chapter 2

One-Dimensional Microscopic Models

This chapter is restricted to one-dimensional motions: pedestrians are assumed to walk on a line toward a common direction. Section 2.1 is dedicated to the so-called Follow-the-Leader model, which relies on the assumption that the instantaneous velocity of an individual is a function of the distance to the next individual.

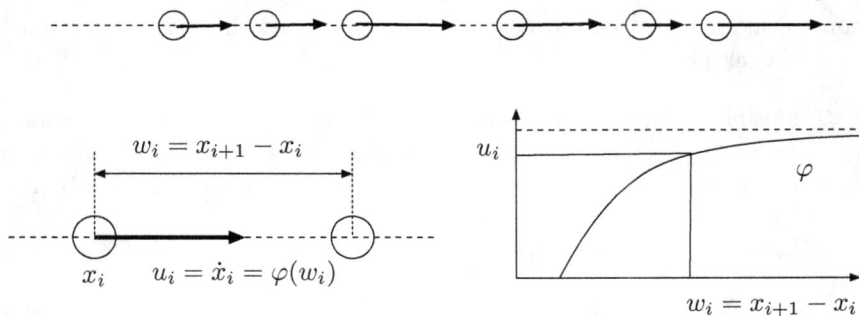

Section 2.2 is dedicated to a richer version of this model, including delayed reaction or inertial effects.

2.1.   Follow-the-Leader Model

We present here the basic form of the Follow-the-Leader (FTL) model, some extensions which have been proposed, and we address some theoretical issues.

FTL model: assumptions and actual behavior

We consider $N + 1$ individuals walking on a straight line. Their respective positions are denoted by

$$x_1(t) < x_2(t) < \cdots < x_{N+1}(t). \tag{2.1}$$

The model is based on the assumption that the instantaneous velocity of i depends upon $x_{i+1} - x_i$ only.

Model 2.1 (FTL model). Let $\varphi : \mathbb{R}_+ \to \mathbb{R}_+$ be a function which assigns a speed to any non-negative distance. The models read:

$$\frac{dx_i}{dt} = \varphi(x_{i+1} - x_i), \quad 1 \leq i \leq N. \tag{2.2}$$

We shall designate by *linear* this model when the speed of individual $N + 1$ is prescribed, and *periodic* the case where $N + 1$ is identified to 1 (in which case the length of the periodic path has to be specified). The system is then autonomous in the latter situation (no explicit dependence upon time), and non-autonomous in the linear case (because $x_{N+1}(t)$ has to be prescribed in the latter situation).

Denoting by $w_m > 0$ (m for "minimal") the inter-individual distance which corresponds to full packing (people are in contact), it is natural to assume that $w \mapsto \varphi(w)$ vanishes at w_m, increases with w, and tends to a maximal value U when w becomes large.

An example of such a function is

$$\varphi(w) = U(1 - \exp(-(w - w_m)/w_s)) \quad \text{for} \;\; w \geq w_m, \tag{2.3}$$

and $\varphi(w) = 0$ otherwise, where w_s is a typical distance below which the modification of the velocity is significant (see Fig. 2.1), and U is the desired speed. Typical values are $U = 1.25\,\mathrm{m\,s}^{-1}$, $w_m = 0.3\,\mathrm{m}$, $w_s = 0.9\,\mathrm{m}$ (see Section 9.4 and Fig. 9.2 therein).

Remark 2.1. Following the literature, we adopt the term "Follow-the-Leader" proposed in Argall *et al.* (2002) to designate this approach. In the

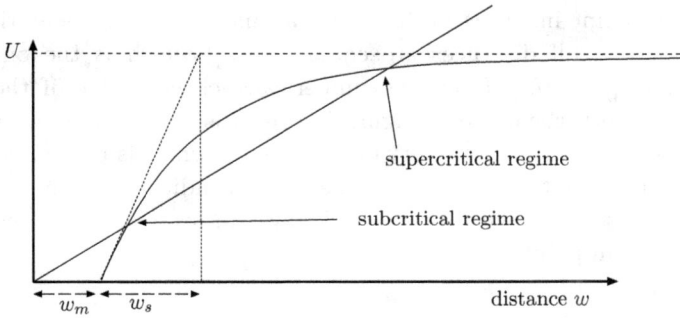

Fig. 2.1.   Speed as a function of the distance.

present context, the term is somewhat improper, since the model is rather based on individuals who have a personal objective (that is moving ahead at velocity U), and who decide to reduce their speed to keep a reasonable distance from the person in front of them, in order to avoid collision. From this standpoint each agent is not really *led* by the agent ahead, but rather *disturbed* or *inhibited*.

Under reasonable assumptions on the behavior function φ, it can be proved that the problem is well-posed in the linear and periodic situations, which means that, for given initial conditions (and prescribed velocity of $N+1$ in the linear case), existence of a unique solution can be established. To be more precise, if the derivative of φ at w_m is not infinite, all distances remain positive for all times (no accident). We put off until the end of this section a rigorous account of those properties.

In order to investigate the stationary states of the system, it is convenient to express it in terms of distances (which can be expected to be constant, whereas positions always move forward). We introduce $w_i = x_{i+1} - x_i$ and, since we shall be interested in the situation where the velocity of $N+1$ is constant and equal to U_{eq}, we define w_{N+1} as w_{eq}. The system now writes

$$\frac{dw_i}{dt} = \varphi(w_{i+1}) - \varphi(w_i), \quad 1 \leq i \leq N. \tag{2.4}$$

Equilibrium point, and stability

Let φ be defined by (2.3). In the linear setting, we prescribe that the speed of $N+1$ is a constant, equal to $U_{\text{eq}} \in]0, U[$. The system then admits an equilibrium point in terms of distances: all individuals move at the same speed U_{eq}, and all distances are equal to w_{eq}, which is the only solution to $\varphi(w_{eq}) = U_{eq}$. It can be numerically checked that if the initial condition is perturbed, the configuration returns to the equilibrium. This expresses the asymptotic stability of the system, which is proven locally by Proposition 2.5. Actually, whatever the initial condition, it can be checked (and proved, see Corollary 2.1) that the system systematically returns to this equilibrium point.

Forced stop and go and upstream propagation

The spontaneous upstream propagation can be observed in the following situation. We consider again the linear setting, and a behavior function defined by (2.3). Starting from a stationary evolution (uniform speed U_{eq}, equal distances w_{eq}), we suppose that $N+1$ suddenly stops, stays still for a while, and then suddenly resume motion at speed U_{eq}. Fig. 2.2 represents the evolution in time of the distances $x_{N+1} - x_N$, $x_N - x_{N-1}, \dots$. Numerical solution is based on the explicit Euler scheme (A.2). The figure shows that the perturbation (stop and go forcing of the head pedestrian) propagates upstream while undergoing dispersion and damping (the profile is stretched and the amplitude decreases with the distance to the head). We refer to Remark 2.6, page 24, for more details on this effect.

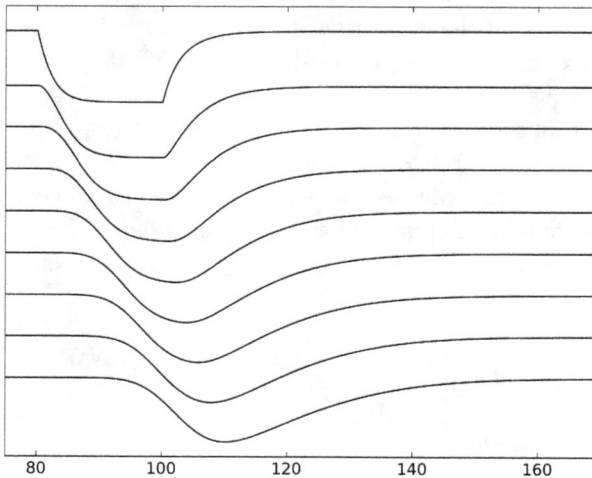

Fig. 2.2.   Distance vs. time (head distance on the top).

Upstream propagation: underlying transport equation

We propose here an informal explanation of the tendency of the system to backward propagate perturbations, as illustrated by Fig. 2.2. A rigorous justification is proposed in Section 2.1.

We consider the equilibrium state in the case where the head entity moves at constant speed $U_{\text{eq}} \in]0, U[$. All pedestrians move at the same speed, and the common distance is such that $U_{\text{eq}} = \varphi(w_{\text{eq}})$. In other words, $W = (w_{\text{eq}}, \ldots, w_{\text{eq}})$ is a stationary solution to the system (2.4). Consider a perturbation of this equilibrium state: we assume that distances are equal to $w_{\text{eq}} + h_i$, where h_i is small with respect to w_{eq}. From (2.4), we have that

$$\dot{h}_i = \varphi(w_{\text{eq}} + h_{i+1}) - \varphi(w_{\text{eq}} + h_i)$$

$$\approx \varphi'(w_{\text{eq}})(h_{i+1} - h_i) = w_{\text{eq}}\varphi'(w_{\text{eq}})\frac{h_{i+1} - h_i}{w_{\text{eq}}}.$$

Since the quantities h_i are defined at points which are roughly equidistributed with a space step equal to w_{eq}, we can interpret the last quotient as the discretization of the space derivative of a function $h(x, t)$, which leads (by anti-discretization) to the equation

$$\frac{\partial h}{\partial t} - \underbrace{w_{\text{eq}}\varphi'(w_{\text{eq}})}_{c}\frac{\partial h}{\partial x} = 0.$$

It is a transport equation with constant velocity $-c$, with $c = w_{\text{eq}}\varphi'(w_{\text{eq}})$, for which solutions are trivially defined as $h(x, t) = H(x + ct)$, where H is any smooth function. Thus, it expresses a transport phenomenon along the array of individuals, in the upstream direction. The velocity c is estimated within a referential which globally moves forward at speed $\varphi(w_{\text{eq}})$. Note that we shall have effective backward propagation in the fixed referential whenever the speed of the wave is larger than the physical speed of entities, i.e. whenever

$$w_{\text{eq}}\varphi'(w_{\text{eq}}) > \varphi(w_{\text{eq}}) \iff \varphi'(w_{\text{eq}}) > \frac{\varphi(w_{\text{eq}})}{w_{\text{eq}}}.$$

In the case where we neglected the size of entities (i.e. $w_m = 0$) the function φ (defined for instance by (2.3), with $w_m = 0$) is typically concave and vanishes at 0. Thus, any chord intersects the graph at a unique point, and the slope of the graph is smaller than the slope of the chord, i.e. $\varphi'(w_{\text{eq}}) < \frac{\varphi(w_{\text{eq}})}{w_{\text{eq}}}$. In this situation, the signal does not go fast enough to actually move upstream in the fixed referential. Conversely, if the size of entities is accounted for (see Fig. 2.1, with $w_m > 0$), we have two regimes for each chord. The first one corresponds to a large density (small distances) and a small physical velocity, so that wave propagates faster than the velocity of entities, which results in an effective backward propagation of information in the fixed referential. In analogy with hydraulic phenomena, this regime can be called *subcritical*. The other regime (supercritical) corresponds to a small density and a large physical velocity. Note that, for a given chord, both regime corresponds to the same pedestrian flow rate $\varphi(w_{eq})/w_{eq}$ (that is the velocity times the density). From a modeling standpoint, the supercritical regime is characterized by the fact that perturbations do not migrate upstream, which makes it easy to model a portion of a larger system. Indeed, if one considers a large collection of individuals moving rightward on a straight line, and one aims at modeling the system restricted to a fixed space interval, it is sufficient to prescribe conditions at the left end of the interval, where people enter the chosen window. In the subcritical regime, since information is likely to enter the domain through the right end, such an approach is inapplicable, and the problem of modeling a fixed portion of the path is much more difficult.

FTL model: mathematical issues

We address in this section mathematical issue pertaining to Model 2.1, i.e. the system of differential equations

$$\frac{dx_i}{dt} = \varphi(x_{i+1} - x_i), \quad 1 \le i \le N, \tag{2.5}$$

with prescribed initial conditions

$$x_1^0 < x_2^0 < \cdots < x_N^0 < x_{N+1}^0,$$

and $\dot{x}_{N+1}(t) = V(t)$ prescribed (or, in the periodic case $N + 1 \equiv 1$).

Well-posedness

Let us start by a "lazy" existence theorem, lazy in the sense that we first disregard the issue of contacts by extending φ on $] - \infty, 0[$ (which does not make much sense from a modeling standpoint). The crucial question of actual contact is put off until the next proposition.

Proposition 2.2. *We consider $N + 1$ individuals, initially located at*

$$x_1^0, \ldots, x_{N+1}^0,$$

such that the order relation (2.1) *is verified. We assume that the velocity of the head entity* $N+1$ *is given as a continuous function of time, taking values in* $[0, U]$. *We assume that the behavior function* φ *is bounded and globally Lipschitzian, with* $\varphi(0) = 0$ (*and identically* 0 *for negative distances*), *with values in* $[0, U]$. *The system* (2.5) *admits a unique maximal solution, which is global.*

Proof. The function φ (extended by 0 on $(-\infty, 0)$) is globally Lipschitzian. Cauchy–Lipschitz theorem (Theorem A.1) applied to $[0, +\infty[\times \mathbb{R}^n$ ensures the existence of a unique maximal solution. Since φ is bounded, this solution is global (see Proposition A.2). $\qquad\square$

Proposition 2.3. *We assume that* φ *vanishes over* $w_m \in [0, +\infty)$, *and that all initial distances are larger than* w_m. *Then, for the solution defined by Proposition 2.2, it holds that* $x_{i+1}(t) - x_i(t) > w_m$ *for all* $t \in [0, +\infty)$.

Proof. We define $L = \|\varphi'\|_\infty$. Let $t \mapsto x(t) \in \mathbb{R}^{N+1}$ be the solution defined by Proposition 2.2. As far as $x_{N+1} - x_N > 0$, it holds that

$$\dot{x}_N = \varphi(x_{N+1} - x_N) \leq L(x_{N+1} - x_N),$$

thus, introducing $w_N = x_{N+1} - x_N$,

$$\dot{w}_N \geq -Lw_N + V(t) \geq -Lw_N,$$

which yields $w_N \geq w_N(0)e^{-Lt}$, which remains positive at all times. All distances w_{N-1}, w_{N-2} can be proved to remain positive in a similar way, by downward induction on indices. $\qquad\square$

Remark 2.2. The Lipschitz character of φ is essential to avoid collision. Consider a function φ which behaves like w^α at 0^+, with $\alpha \in]0, 1[$. Assume that the head entity is staying still at $a \in \mathbb{R}$. The evolution follows

$$\dot{x} = (a - x)^\alpha, \quad x(0) < a,$$

which leads to

$$x(t) = a - \left((a - x(0))^{1-\alpha} - (1 - \alpha)t\right)^{1/1-\alpha}.$$

We then have a collision ($x = a$) in finite time.

Equilibrium points and their stability

Proposition 2.4. *We assume the behavior function* φ *to be defined by* (2.3). *Let the speed of* $N+1$ *be set to a constant value* $U_{\text{eq}} \in (0, U)$, *and* w_{N+1} *be defined as* $\varphi^{-1}(U_{\text{eq}})$. *Then the system* (2.4) *admits a unique equilibrium point* $W = (w_{\text{eq}}, \ldots, w_{\text{eq}})$.

Remark 2.3. The previous proposition straightforwardly extends to behavior functions which map $[w_m, +\infty)$ onto $[0, U)$ increasingly.

Proposition 2.5. *The equilibrium point of Proposition 2.4 is asymptotically stable.*

Proof. System (2.4) writes $\dot{W} = F(W)$. We have $F(W_{\text{eq}}) = 0$, and

$$\nabla F(W_{\text{eq}}) = \varphi'(w_{\text{eq}}) \begin{pmatrix} -1 & 1 & 0 & \cdot & 0 \\ 0 & -1 & 1 & \cdot & \cdot \\ \cdot & \cdot & \cdot & \cdot & 0 \\ \cdot & \cdot & \cdot & -1 & 1 \\ 0 & \cdot & \cdot & 0 & -1 \end{pmatrix}. \tag{2.6}$$

This matrix has a unique eigenvalue $-\varphi'(w_{\text{eq}}) < 0$, which proves the asymptotic stability with a characteristic time equal to $\eta = 1/\varphi'(w_{\text{eq}})$. $\qquad\square$

We shall see that, in the present case of a non diagonalizable matrix, the effective characteristic time may be significantly larger than $1/\varphi'(w_{\text{eq}})$. More precisely, this time is not uniform with respect to the number N of individuals, whereas $1/\varphi'(w_{\text{eq}})$ is (see Remark 2.5, page 23, on the extreme case of an infinite number of individuals).

In the periodic setting, we consider a corridor of length L. The equations write

$$\dot{x}_i = \varphi(x_{i+1} - x_i), \quad i = 1, \ldots, N \quad (N+1 \equiv 1),$$

or, expressed in terms of distances $w_i = x_{i+1} - x_i$,

$$\dot{w}_i = \varphi(w_{i+1}) - \varphi(w_i), \quad i = 1, \ldots, N \quad (N+1 \equiv 1), \tag{2.7}$$

which we write $\dot{w} = F(w)$.

If φ is increasing, the system (2.7) obviously admits a unique equilibrium point $w_{\text{eq}} = (L/N, \ldots, L/N)$.

Proposition 2.6. *Assume that φ is a C^1 function on $[0, +\infty[$, with φ' positive. The equilibrium point $w_{\mathrm{eq}} = (L/N, \ldots, L/N)$ is asymptotically stable.*

Proof. We write the gradient of F at w_{eq}:

$$\nabla F(w_{\mathrm{eq}}) = \mathbf{M} = \varphi'(w_{\mathrm{eq}}) \begin{pmatrix} -1 & 1 & 0 & \cdot & 0 \\ 0 & -1 & 1 & \cdot & \cdot \\ \cdot & \cdot & \cdot & \cdot & 0 \\ \cdot & \cdot & \cdot & -1 & 1 \\ 1 & \cdot & \cdot & 0 & -1 \end{pmatrix}$$

$$= \beta \mathbf{A}_{\mathrm{per}} = \beta \left(-\mathbf{I} + \mathbf{C} \right), \tag{2.8}$$

with $\beta = \varphi'(w_{\mathrm{eq}})$, and $\mathbf{C}$ is a circulant matrix. It verifies $\mathbf{C}^N = \mathbf{I}$ and the family $(\mathbf{C}^k)_{0 \leq k \leq N-1}$ is linearly independent. Its characteristic polynomial is therefore $X^N - 1$, and its eigenvalues are the N roots of unity. The eigenvalues of $\mathbf{A}_{\mathrm{per}} = -\mathbf{I} + \mathbf{C}$ are then

$$\mu_k = -1 + \exp\left(\frac{2ik\pi}{N} \right), \quad k = 0, \ldots, N-1, \tag{2.9}$$

and the eigenvalues of $\mathbf{M}$ are those μ_k's multiplied by $\beta = \varphi'(w_{\mathrm{eq}}) > 0$. Thus, all eigenvalues of $\mathbf{M}$ have a non-positive real part, which suggests stability of the system. Yet, for $k = 0$, we have $\mu_0 = 0$, so that Theorem A.2 does not ensure asymptotic stability. This asymptotic stability can nevertheless be proved by noticing that the associated eigenspace is $\mathbb{R}e$, where e is the vector with all elements equal to 1. Since, by construction, the sum of the w_i's is constant (equal to L), the admissible perturbations have a zero mean, thus they are orthogonal to e. Besides, we immediately check that $e^\perp$ is stable through A_{per}. We can therefore restrict the stability investigation to $e^\perp$, in which all eigenvalues have a negative real part.[1]   □

---

[1] We may also follow a more standard approach by eliminating a redundant variable, for example by writing $w_N = L - \sum_{i=1}^{N-1} w_i$. The last equation then writes $w_{N-1} = \varphi(L - \sum_{i=1}^{N-1} w_i) - \varphi(w_{N-1})$, and the gradient writes

$$\nabla F(w_{\mathrm{eq}}) = \varphi'(w_{\mathrm{eq}}) \begin{pmatrix} -1 & 1 & 0 & \cdot & 0 \\ 0 & -1 & 1 & \cdot & \cdot \\ \cdot & \cdot & \cdot & \cdot & 0 \\ \cdot & \cdot & \cdot & -1 & 1 \\ -1 & -1 & \cdot & -1 & -1 \end{pmatrix}.$$

Remark 2.4 (Relaxation time). The real part with smallest absolute value is $\varphi'(w_{\text{eq}})(1 - \cos(2\pi/N))$, which is close to $\varphi'(w_{\text{eq}})2\pi^2/N^2$. As a consequence, the characteristic time is of the order

$$\tau \approx \frac{1}{2\pi^2}\frac{N^2}{\varphi'(w_{\text{eq}})}.$$

This relaxation occurs along a mode of *low frequency* in space.

Rigorous justification of upstream propagation

The upstream propagation of perturbations described in the previous section p. 17 is a consequence of the particular form of the matrix M given by (2.6). The linearized system writes

$$\frac{dw}{dt} = \mathbf{M}w.$$

We keep the notation w to designate the vector of unknowns, but the w_i now correspond to variations around the equilibrium point. The solution to the linear problem above writes

$$w(t) = e^{t\mathbf{M}}w^0,$$

where w^0 is an initial perturbation. The matrix $\mathbf{M}$ can be written

$$\mathbf{M} = \beta(-\mathbf{I} + \mathbf{N}) \qquad (2.10)$$

with $\beta = \varphi'(w_{\text{eq}})$, and $\mathbf{N}$ is the nilpotent matrix

$$\mathbf{N} = \begin{pmatrix} 0 & 1 & 0 & \cdot & 0 \\ 0 & 0 & 1 & \cdot & \cdot \\ \cdot & \cdot & \cdot & \cdot & 0 \\ \cdot & \cdot & \cdot & 0 & 1 \\ 0 & \cdot & \cdot & 0 & 0 \end{pmatrix}, \quad \mathbf{N}^2 = \begin{pmatrix} 0 & 0 & 1 & \cdot & 0 \\ 0 & 0 & 0 & \cdot & \cdot \\ \cdot & \cdot & \cdot & \cdot & 1 \\ \cdot & \cdot & \cdot & 0 & 0 \\ 0 & \cdot & \cdot & 0 & 0 \end{pmatrix}, \ldots, \mathbf{N}^N = 0.$$

Because of the nilpotent character of $\mathbf{N}$, the exponential takes the following polynomial expression:

$$e^{t\mathbf{M}} = e^{-\beta t}\left(\mathbf{I} + \beta t\mathbf{N} + \frac{(\beta t)^2}{2!}\mathbf{N}^2 + \cdots + \frac{(\beta t)^{N-1}}{(N-1)!}\mathbf{N}^{N-1}\right). \qquad (2.11)$$

The characteristic polynomial P_{N-1} of this matrix verifies $P_{N-1} = -\lambda P_{N-2} + (-1)^N$, so that

$$P_{N-1} = (-1)^{N+1}(1 + \lambda + \cdots + \lambda^{N-1}),$$

which shows that the eigenvalues are the non-trivial roots of unity.

Let us show that this very form explains the backward propagation of perturbations. Consider a perturbation concentrated at the head of the line, i.e. colinear to $w^0 = e_N$, where e_i is the ith vector of the canonical basis of $\mathbb{R}^N$. It holds that

$$\mathbf{N}e_N = e_{N-1}, \mathbf{N}^2 e_N = e_{N-2}, \ldots, \mathbf{N}^{N-1} e_N = e_1.$$

The general behavior of the solution to the linearized system can therefore be expressed in terms of perturbations experienced by all individuals in the line. More precisely, the individual of index k undergoes the perturbation

$$\frac{(\beta t)^{N-k}}{(N-k)!} e^{-\beta t}, \quad k = 1, \ldots, N.$$

To estimate the velocity associated to this propagation phenomenon, let us compute at which instant the perturbation undergone by k is maximal:

$$p_k(t) = e^{-\beta t} \frac{(\beta t)^k}{k!}, \quad p_k'(t) = e^{-\beta k} \frac{\beta^k t^{k-1}}{k!} (-\beta t + k), \tag{2.12}$$

which vanishes for $t = k/\beta$. Since the distance between entities is of the order w_{eq}, it reflects an upstream propagation at velocity

$$c = -\beta w_{\mathrm{eq}} = -w_{\mathrm{eq}} \varphi'(w_{\mathrm{eq}}).$$

Remark 2.5. To apprehend what happens when the number of individuals is large, let us consider the case of an infinite number of pedestrians, moving at a constant speed behind a leader (or more properly an inhibitor, see Remark 2.1), the velocity of which is prescribed. At some instant $t > 0$, the collection of perturbations is (pedestrians are now indexed starting from 0 for the leader, up to infinity)

$$p(t) = (p_k(t))_{k \in \mathbb{N}}, \quad p_k = e^{-\beta t} \frac{(\beta t)^k}{k!},$$

that is a Poisson distribution with parameter βt, with $\beta = \varphi'(w_{\mathrm{eq}})$. Note that

$$\|p(t)\|_1 = \sum_{k=0}^{\infty} e^{-\beta t} \frac{(\beta t)^k}{k!} = 1,$$

i.e. the total mass of the perturbation remains constant. There is no asymptotic stability for this norm, in the case of an infinite number of pedestrians.

The perturbation nevertheless vanishes in some way, if one considers for instance the uniform norm $\|p(t)\|_\infty = \max_{k \in \mathbb{N}} p_k(t)$.

Remark 2.6. The dispersion/damping of the propagation represented in Fig. 2.2 can be explained by the previous development. Indeed, from (2.12), we obtain the expression of the maximal value of the perturbation for individual $N - k$. The maximum is attained at $t = k/\beta$, and it writes

$$p_k^{\max} = e^{-k} \frac{k^k}{k!}.$$

From Stirling's formula, this quantity is equivalent to $1/\sqrt{2\pi k}$ for large values of k, which expressed a slow decrease of the perturbation amplitude downstream the line.

A deeper look on upstream propagation in the periodic case

In the non-periodic case, the particular form of the exponential of the linearized system made it possible to establish the upstream propagation of perturbations at velocity $-w_{\mathrm{eq}}\varphi'(w_{\mathrm{eq}})$. To that purpose, we considered the case of a perturbation localized at the head to the line. We propose here to quantify this propagation phenomenon in the periodic situation. We consider a circular corridor of length L, along which N pedestrians walk. The equilibrium state corresponds to equidistributed distances (common distance $w_{\mathrm{eq}} = L/N$), with all individuals moving at speed $\varphi(w_{\mathrm{eq}})$. From (2.8), the linearized system writes

$$\frac{dw}{dt} = \mathbf{M}w = \varphi'(w_{\mathrm{eq}})\mathbf{A}_{\mathrm{per}}w = \varphi'(w_{\mathrm{eq}})\left(-\mathbf{I} + \mathbf{C}\right)w.$$

As already noticed (see (2.9)), the matrix $\mathbf{M}$ is diagonalizable, with eigenvalues

$$\beta\mu_k = \beta\left(-1 + \exp\left(\frac{2ik\pi}{N}\right)\right) \in \mathbb{C}, \quad \beta = \varphi'(w_{\mathrm{eq}})$$

and associated eigenvectors

$$v_k = \left(\exp\left(\frac{2ik\pi m}{N}\right)\right)_m \in \mathbb{C}^N.$$

For such a linear system, the solution associated to the initial condition w^0 is $w(t) = e^{t\mathbf{M}}w^0$. If one consider an initial perturbation along the kth eigenmode, with $k = 1, \ldots, N - 1$, we obtain

$$w(t) = \underbrace{\exp(t\beta \operatorname{Re}(\mu_k))}_{\text{damping}} \underbrace{\exp(it\beta \operatorname{Im}(\mu_k))v_k}_{\text{propagation}}, \tag{2.13}$$

with the standard decomposition $\mu_k = \operatorname{Re}(\mu_k) + i \operatorname{Im}(\mu_k)$. The first factor indeed encodes a damping effect, since real parts are negative:

$$\operatorname{Re}(\mu_k) = -\beta\left(1 - \cos\left(\frac{2k\pi}{N}\right)\right) < 0, \quad \forall k = 1, \ldots, N - 1.$$

The propagation in space is encoded in the second factor of (2.13). Indeed, the ℓth entry of $\exp(it\beta \operatorname{Im}(\mu_k))v_k$ expresses

$$\exp\left(i\beta \sin\left(\frac{2k\pi}{N}\right)t\right)\exp\left(\frac{2ik\pi\ell}{N}\right) = \exp\left(\frac{2ik\pi}{N}\left(\ell + \underbrace{\frac{\beta N}{2\pi k}\sin\left(\frac{2k\pi}{N}\right)}_{=c_k}t\right)\right)$$

$$= \exp\left(\frac{2ik\pi}{N}(\ell + c_k t)\right).$$

Consider now the vector $W(t)$, the ℓth element of which is the complex number above. At time $t + \tau$, with $\tau = 1/c_k$, the corresponding vector $W(t+\tau)$ has undergone a 1-step leftward shift, i.e. the ℓth entry of $W(t+\tau)$ is the $(\ell + 1)$th entry of $W(t)$. This expression therefore corresponds to a leftward propagation at constant speed c_k (expressed in pedestrian per second). Note that we recover, for k/N small, a speed of the order of $-\beta = -\varphi'(w_{\text{eq}})$ (expressed in s^{-1}, or *entities* per second). If one accounts for the fact that pedestrians are separated by a distance close to w_{eq}, we recover the metric velocity $-w_{\text{eq}}\varphi'(w_{\text{eq}})$ (in m s^{-1}).

We also note a *dispersion* phenomenon: the propagation of a perturbation depends on its wavelength. For instance, for the space frequency which will play a crucial role in the inertial version of the model, that is for $k = N/6$, we have a slightly slower speed (factor $3/\pi \sin(\pi/3)$).

Global stability

The stability analysis above relies on the linearized problem around the equilibrium. Thus, it only guarantees *local* asymptotic stability (in the sense of Definition A.3. We prove here that the asymptotic stability is actually global as soon as the behavior function φ is (strictly) increasing. We shall limit ourselves here to the periodic case[2], which makes it possible to explicitly build a Lyapunov functional.

Proposition 2.7. *We consider Model 2.1 in the periodic setting (pedestrians in a circular corridor), i.e. $N + 1$ is identified with 1. We assume that φ is C^1 on $[0, +\infty)$ a non-decreasing. We denote $w_i = x_{i+1} - x_i$, and we consider a solution to the associated system*

$$\frac{dw_i}{dt} = \varphi(w_{i+1}) - \varphi(w_i), \quad 1 \leq i \leq N. \tag{2.14}$$

For any convex function $g : [0, +\infty) \to \mathbb{R}$, the quantity

$$S(w(t)) = \sum_i g(w_i)$$

is non-increasing.

If one assumes that φ is (strictly) increasing, and that g is strictly convex, then $t \mapsto S(w(t))$ strictly decreases unless w is the equilibrium point (i.e. all distances are equal to L/N).

Proof. We have that

$$\frac{d}{dt}\left(\sum_i g(w_i)\right) = \sum_i g'(w_i)\frac{dw_i}{dt} = \sum_i g'(w_i)\left(\varphi(w_{i+1}) - \varphi(w_i)\right)$$

$$= \sum_i \varphi(w_i)(g'(w_{i-1}) - g'(w_i)).$$

Suppose now that g is strictly convex. The function g' is then increasing, and one can make the change of variable $\gamma_i = g'(w_i)$. The quantity above expresses

$$\sum_i \varphi \circ (g')^{-1}(\gamma_i)(\gamma_{i-1} - \gamma_i),$$

where $\varphi \circ (g')^{-1}$ is increasing. It therefore writes as the derivative of a convex function: $\varphi \circ (g')^{-1}(\gamma) = \psi'(\gamma)$. Since ψ is convex, it holds that

$$\psi(\gamma_i) + \psi'(\gamma_i)(\gamma_{i-1} - \gamma_i) \leq \psi(\gamma_{i-1}), \tag{2.15}$$

---

[2]The property is also valid in the non-periodic situation, but it relies on alternative arguments.

so that

$$\frac{d}{dt}\left(\sum_i g(w_i)\right) \leq \sum_i (\psi(\gamma_{i-1}) - \psi(\gamma_i)) = 0.$$

If g is not strictly convex, we apply the previous procedure to the strictly convex function $g_\varepsilon(z) = g(z) + \varepsilon z^2$, and then have ε go to 0.

Now if g is strictly convex, and φ is increasing, then inequality (2.15) is strict for at least one i unless all β_i's are identical. □

Corollary 2.1. *Under the assumptions of Proposition 2.7, in the strict setting (i.e. φ is strictly increasing), the unique equilibrium point is globally attractive, i.e. all solutions converge to it, for any initial condition.*

Proof. For any strictly convex function g (e.g. $g(x) = x\log(x)$), the function $S(w) = \sum_i g(w_i)$ is a strict Lyapunov functional, hence the global asymptotic stability by Theorem A.2. □

Remark 2.7. In the case where the length is equal to 1, one may interpret $w = (w_i)$ as a probability measure on a discrete set with N elements. Taking $g(x) = x\log x$, we proved that the *entropy* of this probability law

$$S(w) = \sum_i w_i \log w_i$$

decreases.

Critical discussion on the FTL model

The FTL model rests on the assumption that individuals instantaneously adopt a velocity according to the current distance which they estimate. In reality, this process is not instantaneous, for two main reasons: psychological delays and mechanical considerations (inertia). The instantaneousness of this velocity control prevents this model from recovering spontaneous instabilities which are observed in some situations. Those issues are addressed in Section 2.2.

Besides, the model in its native form does not account for variability of behaviors. This is a very general issue pertaining to the modeling of many-body systems, an issue that is particularly sensitive in case of living entities like human beings. The approach we presented was based on the assumption that all individuals have the very same behavior. The choice makes it possible to design elegant models with very few parameters, which lend themselves well to mathematical analysis. Moreover, as pointed out in Helbing *et al.* (2000): "*Although, in reality, most parameters are varying individually, we chose identical values for all pedestrians to minimize the number of parameters for reasons of calibration and robustness.*" It can be expected that such an unrealistic assumption will not harm the general features exhibited by the model. Yet, it is important to establish some sort of structural stability of the model, i.e. to verify that slightly modifying the assumptions is not likely to deeply affect the main features of the model. We propose to perform such an investigation in Model 2.1. We shall see that some mathematical properties are indeed robust with respect to perturbations on the individual behaviors. Yet, in the present situation, it will be emphasized that the core argument to properly establish the upstream propagation of perturbations has been based on the very characteristic of the matrix **M** encoding the linearized problem (see Section 2.1). More precisely, the expression (2.10), which is the basis of further developments on upstream propagation, expresses that the matrix **M** is not diagonalizable, which is clearly problematic in terms of robustness, since this property is *not generic* for matrices.[3]

Consider Model 2.1, with a head individual $N+1$ moving at speed $U_{\text{eq}} = \varphi_{N+1}(w_{\text{eq}})$, and assume that each individual i has their own behavior function φ_i. If we assume that all those functions are increasing for $w \geq w_m$, we can establish existence and uniqueness of a unique equilibrium point as soon as the head velocity is attainable by followers, i.e.

$$U_{\text{eq}} < \sup_w \varphi_i(w) \quad \forall i.$$

Let us define w_{eq}^i as the distance which realizes U_{eq}, i.e. such that $U_{\text{eq}} = \varphi_i(w_{\text{eq}}^i)$. The vector $w_{\text{eq}}^1, \ldots, w_{\text{eq}}^N$ is then the unique equilibrium point. One may say that Proposition 2.4 is stable with respect to behavior variability.

The stability analysis of this equilibrium point relies on the matrix

$$\nabla F = \begin{pmatrix} -\beta_1 & \beta_2 & 0 & & \cdot & 0 \\ 0 & -\beta_2 & \beta_3 & & \cdot & \cdot \\ \cdot & \cdot & \cdot & & \cdot & 0 \\ \cdot & & \cdot & -\beta_{N-1} & \beta_N \\ 0 & & \cdot & \cdot & 0 & -\beta_N \end{pmatrix}, \quad \beta_i = \varphi_i'(w_{\text{eq}}^i), \quad i = 1, \ldots, N.$$

$$(2.16)$$

---

[3]In the sense that the set of non-diagonalizable matrices has zero Lebesgue measure in $\mathbb{R}^{N \times N}$.

The situation is somewhat puzzling since, if one expects the upstream propagation phenomenon to be reproduced by this new model, the structure is fully different from a mathematical standpoint. Indeed, the parameters β_i have no reason to be identical, and it is reasonable to consider that, while they are likely to be close to each other (small variability), they are generically all distincts. But then the matrix M is diagonalizable, the expression (2.11) is ruled out, and describing $e^{tM} w_0$ is a fully different story.

This raises delicate questions, because the diagonalizable matrix $\mathbf{M}$ is close to the set of non-diagonalizable matrices, more precisely to the set of matrices which have a single, one-dimensional, eigenspace. It implies that the change of basis matrix will be almost singular, i.e. all column vectors are almost colinear. It makes it risky to carry out an actual diagonalization of the matrix, although the eigenvalues are explicitly known.

We may actually check that the model possesses some sort of structural stability without diagonalizing the matrix. We formulate the problem in the following way. Consider the perturbed linear system

$$\frac{dw}{dt} = (\mathbf{M} + \varepsilon \mathbf{B})\, w,$$

where M is the non-diagonalizable matrix involved in the native model, and B the matrix which accounts for individual variability:

$$\mathbf{M} + \varepsilon \mathbf{B} = \begin{pmatrix} -1 & 1 & 0 & \cdot & 0 \\ 0 & -1 & 1 & \cdot & \cdot \\ \cdot & \cdot & \cdot & \cdot & 0 \\ \cdot & \cdot & \cdot & -1 & 1 \\ 0 & \cdot & \cdot & 0 & -1 \end{pmatrix} + \varepsilon \begin{pmatrix} -b_1 & b_2 & 0 & \cdot & 0 \\ 0 & -b_2 & b_3 & \cdot & \cdot \\ \cdot & \cdot & \cdot & \cdot & 0 \\ \cdot & \cdot & \cdot & -b_{N-1} & b_N \\ 0 & \cdot & \cdot & 0 & -b_N \end{pmatrix}.$$

Denote by w^0 the solution of the reference system $\dot{w}^0 = \mathbf{M} w^0$, and express the solution to the perturbed system in the form $w_\varepsilon = w^0 + \varepsilon w^1$. We obtain

$$\dot{w}^1 = \mathbf{M} w^1 + \mathbf{B} w^0 + \varepsilon \mathbf{B} w^1,$$

which converges toward a bounded solution when ε goes to 0. It ensures that the behavior of the perturbed system is close to the one associated to the reference system, with a first order correction

$$w^1(t) = \int_0^t e^{-(t-s)\mathbf{M}} B w^0(s)\, ds,$$

which reflects the effects due to the disparities between individuals.

2.2.   Accounting for Inertia/Delays

The section is dedicated to an inertial version of the FTL model. We also present here an extension of the FTL model obtained by including some time delay effect. As we shall see, both models are very similar, in particular their linearized versions are identical. In the inertial setting, this model can be seen as a one-dimensional version of the Social Force model (Helbing *et al.*, 1995), with asymmetric forcing terms to account for the fact that pedestrians are influenced by their immediate neighbor in front of them (see Chapter 3). A similar model is investigated in Chraibi *et al.* (2015).

Model 2.2 (Inertial FTL model). Like in Section 2.1, we consider $N+1$ individuals walking on a straight line. Their respective positions are denoted by

$$x_1(t) < x_2(t) < \cdots < x_{N+1}(t). \qquad (2.17)$$

The model is based on a behavior function φ (a typical example is represented in Fig. 2.1). We shall consider that the actual velocity of an individual relaxes toward the velocity that is associated to the current distance from the person in front of them, with a characteristic time τ. The model therefore reads

$$\ddot{x}_i = \frac{1}{\tau}(\varphi(x_{i+1} - x_i) - \dot{x}_i). \qquad (2.18)$$

Remark 2.8. This model has a straight mechanical interpretation: each individual is considered as an entity of mass m, which instantaneously measures the distance to the entity in front, computes the velocity which it considers adapted to this distance (function φ), and exerts a force on itself in order to approach this velocity. In this setting, the force is simply

$$\frac{m}{\tau}(\varphi(x_{i+1} - x_i) - \dot{x}_i).$$

Remark 2.9. The model (2.18) can be seen as a mono-dimensional version of the Social Force model presented in Section 3.1. Indeed, if one assumes that φ is given by (2.3), it writes

$$\ddot{x}_i = \frac{1}{\tau}(\varphi(x_{i+1} - x_i) - \dot{x}_i)$$

$$= \frac{1}{\tau}(U - \dot{x}_i) - \frac{U}{\tau}\exp(-(x_{i+1} - x_i - w_m)/w_s).$$

- *FTL model with delay*

We present here an alternative approach based on different assumptions: each individual is able to instantaneously control their velocity, but this velocity is based on a distance which is not the current one. We rather have velocity depend on some sort of "psychological distance" which tends to relax to the real one with some delay τ:

$$\frac{dx_i}{dt} = \varphi(\tilde{w}_i), \tag{2.19}$$

$$\frac{d\tilde{w}_i}{dt} = \frac{1}{\tau}\left(x_{i+1} - x_i - \tilde{w}_i\right). \tag{2.20}$$

Remark 2.10. The delay is implemented here in a smooth way, this will make it possible to perform a stability analysis using standard tools. Implementing a delayed model in a strict sense would consist in setting $\tilde{w}_i(t) = w_i(t - \tau)$. Such an approach is proposed in Lemercier *et al.* (2016), where the delay affects the relative velocity between two pedestrians. More generally, such effects can be accounted for by mean of an integration kernel K, that is a non-negative function defined over $\mathbb{R}^+$, with integral equal to one. The formula is then

$$\tilde{w}_i(t) = \int_0^\infty K(s)w_i(t - s)\,ds.$$

Model (2.19)–(2.20) roughly corresponds to an exponential kernel, whereas the pure delayed model would consist in choosing K as a Dirac mass at τ.

Inertial FTL model: mathematical issues

Existence and uniqueness of solutions, collisions

We address in this section mathematical issues pertaining to Model 2.2, which we write as a first-order in time model:

$$\begin{cases} \dot{x}_i = v_i, \\ \dot{v}_i = \frac{1}{\tau}(\varphi(x_{i+1} - x_i) - v_i), \end{cases} \tag{2.21}$$

with prescribed initial conditions

$$x_1^0 < x_2^0 < \cdots < x_N^0 < x_{N+1}^0, \ v_1^0, \ \ldots, \ v_N^0$$

and $\dot{x}_{N+1}(t) = V(t)$ prescribed (or, in the periodic case $N + 1 \equiv 1$).

If one assumes that the function φ is Lipschitzian in $[0, +\infty)$ and $\varphi(0) = 0$, its extension by 0 over $]-\infty, 0]$ remains Lipschitzian, and Cauchy–Lipschitz theorem (Theorem A.1) applied to the system (2.21) ensures the existence of a unique maximal solution, which is global thanks to Proposition A.2. Solutions for which distances vanish or may even become negative are to be considered with a special attention, since they do not correspond to realistic evolutions. Indeed, if one of the distances vanishes, it corresponds to a collision and the model itself, although it may lead to mathematically viable solutions, does not make sense. Let us check that such accidents are indeed likely to happen. We consider the oversimplified situation of an individual moving in the direction of another individual at rest at position 0. The position $x(t) \leq 0$ of the moving person verifies

$$\ddot{x} = \frac{1}{\tau}\left(\varphi(-x) - \dot{x}\right),$$

with initial conditions on position $x(0) = x^0 < 0$ and velocity $v(0) = v^0 > 0$. We are interested on the behavior of the system in the neighborhood of 0, thus $\varphi(-x) \approx -\varphi'(0)\,x$. Denoting $\varphi'(0) = 1/\eta_0$, we obtain

$$\ddot{x} + \frac{1}{\tau}\dot{x} + \frac{1}{\tau\eta_0}x = 0.$$

The roots of the characteristic equation are

$$\lambda = \frac{1}{2\tau}\left(-1 \pm \sqrt{1 - \frac{4\tau}{\eta_0}}\right).$$

We therefore have a non-oscillatory damping as soon as $\tau/\eta_0 < 1/4$. Otherwise, x hits the position 0 (with a non-zero velocity), therefore accidents can not be excluded, and physically viable solutions are defined locally only.

Stability analysis

We investigate here the stability of model (2.21). We shall focus on the periodic case (circular corridor, like in the experiments presented in Lemercier *et al.* (2016) or Portz and Seyfried (2011)). We consider N pedestrians on a periodic corridor of length L. The stationary regime is characterized by equidistant pedestrians, i.e. $x_{i+1} - x_i \equiv w_{\text{eq}} = L/N$, with a common velocity $V = \varphi(w_{\text{eq}})$.

We introduce the distance variables $w_i = x_{i+1} - x_i$. For this new variable the system writes

$$\ddot{w}_i = \frac{1}{\tau}(\varphi(w_{i+1}) - \varphi(w_i) - \dot{w}_i), \qquad (2.22)$$

for which the vector $(w_{eq}, w_{eq}, \dots, w_{eq})$ is an equilibrium point. This model can be written at the first order in time in the variable $y = (\dot{w}, \dot{v}) = \Psi(w, v)$, with $v = \dot{w}$. In this setting, the equilibrium point is

$$y_{eq} = (w_{eq}, \dots, w_{eq}, 0, \dots, 0).$$

The stability of the equilibrium point is determined by the properties of the matrix

$$\nabla\Psi_{|y=y_{eq}} = \begin{pmatrix} 0 & \mathbf{I} \\ \frac{1}{\tau}\varphi'(w_{eq})\mathbf{A}_{per} & -\frac{1}{\tau}\mathbf{I} \end{pmatrix}, \qquad (2.23)$$

with

$$\mathbf{A}_{per} = \begin{pmatrix} -1 & 1 & 0 & \cdot & 0 \\ 0 & -1 & 1 & \cdot & \cdot \\ \cdot & \cdot & \cdot & \cdot & 0 \\ 0 & \cdot & \cdot & -1 & 1 \\ 1 & 0 & \cdot & 0 & -1 \end{pmatrix}. \qquad (2.24)$$

The matrix $\mathbf{A}_{per}$ is the sum of $-\mathbf{I}$ and a circulant matrix $\mathbf{C}$. The latter verifies $\mathbf{C}^N = \mathbf{I}$, and the family $(\mathbf{C}^k)_{0 \le k \le n-1}$ is linearly independent, the characteristic polynomial of $\mathbf{C}$ is therefore $X^N - 1$, thus its eigenvalues are the N roots of unity. The eigenvalues of $\mathbf{A}_{per}$ are

$$\mu_k = -1 + \exp\left(\frac{2ik\pi}{N}\right), \quad k = 0, \dots, N-1.$$

The eigenvalue problem for the global matrix writes

$$v = \lambda w, \quad \frac{\varphi'(w_{eq})}{\tau}\mathbf{A}_{per}w - \frac{1}{\tau}v = \lambda v \implies \left(\lambda^2 + \frac{\lambda}{\tau} - \frac{\varphi'(w_{eq})}{\tau}\mathbf{A}_{per}\right)w = 0.$$

For any eigencouple w_k, $\mu_k = -1 + \exp\left(\frac{2ik\pi}{N}\right)$ of $\mathbf{A}_{per}$, we shall have two eigenvalues for the global matrix, which are the roots of

$$\lambda^2 + \frac{\lambda}{\tau} - \frac{\varphi'(w_{eq})}{\tau}\mu_k = 0,$$

i.e.

$$\lambda_k^{\pm} = \frac{1}{2\tau}\left(-1 \pm \sqrt{1 - 4\varphi'(w_{\text{eq}})\tau\left(1 - \exp\left(\frac{2ik\pi}{N}\right)\right)}\right). \qquad (2.25)$$

Introducing $\alpha = 4\varphi'(w_{\text{eq}})\tau$, the eigenvalues $\lambda_k^{\pm}$ lie on a geometric zone which is the image of the unit circle by the (bivalued) mapping defined in the complex plane by

$$z \longmapsto (-1 \pm \sqrt{1 - \alpha(1 - z)})/2\tau.$$

the crucial point is to determine whether some of those eigenvalues may have a positive real part. The question can be formulated as follows: does the square root of the circle centered at $1 - \alpha$ with radius α belong to the half-plane $\text{Re}(z) \leq 1$?

This question can be answered rigorously:

Lemma 2.1. *The square root of the circle centered at $1 - \alpha$ with radius α intersects the open half-plane $\text{Re}(z) > 1$ if and only if $\alpha > 2$.*

Proof. The set which we aim at describing contains all those $\bar{x} + i\bar{y}$ such that

$$\bar{x}^2 - \bar{y}^2 = x, \quad 2\bar{x}\bar{y} = y$$

where $x + iy$ ranges over the circle of equation $(x - 1 + \alpha)^2 + y^2 = \alpha^2$. It is therefore a *quartic* curve of equation

$$\left(\bar{x}^2 - \bar{y}^2 - 1 + \alpha\right)^2 + 4\bar{x}^2\bar{y}^2 = \alpha^2,$$

which contains $z = 1$.

Let us introduce $X = \bar{x}^2$, $Y = \bar{y}^2$, to obtain

$$\Psi(X, Y) = (X - Y - 1 + \alpha)^2 + 4XY - \alpha^2 = 0.$$

The derivative of Ψ with respect to X, which is $2(X + Y - 1 + \alpha)$ does not vanish at $(1, 0)$. The Implicit Function Theorem makes it possible to express X as a function of Y in the neighborhood of $(1, 0)$, and we have that

$$\frac{dX}{dY}_{|(1,0)} = \frac{2 - \alpha}{\alpha},$$

which is positive (i.e. abscissae go above the value 1) as soon as $\alpha > 2$. $\quad\square$

Remark 2.11. For $\alpha = 1/2$, eigenvalues are located on a *Lemniscate of Bernoulli* (see Fig. 2.3). For $\alpha = 1$, the quartic is the unit circle (in fact

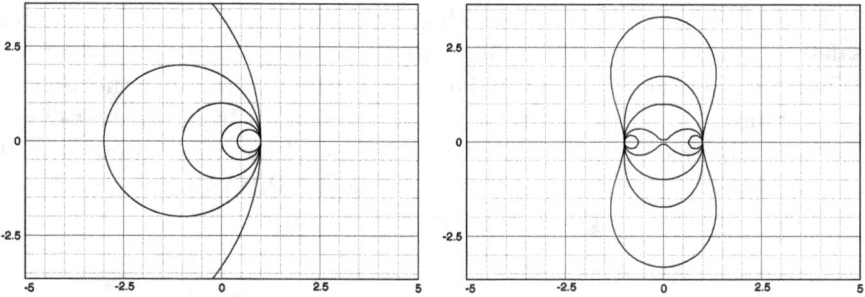

Fig. 2.3. Circles (left) and associated quartic curves (right), for $\alpha = 0.3, 0.5, 1, 2, 6$.

two superposed copies of the unit circle). For the critical value $\alpha = 2$, the shape is that of a stadium, with a zero curvature at 1; for $\alpha > 2$, the set delimits a zone which is no longer convex (biconcave shape).

Remark 2.12. The stability analysis also applies to the model with delay (Eqs. (2.19) and (2.20)) expressed in terms of distances ($w_i = x_{i+1} - x_i$) and psychological distances:

$$\frac{dw_i}{dt} = \varphi(\tilde{w}_{i+1}) - \varphi(\tilde{w}_i),$$

$$\frac{d\tilde{w}_i}{dt} = \frac{1}{\tau}(w_i - \tilde{w}_i).$$

Indeed, the linearized system is characterized by the matrix

$$\begin{pmatrix} 0 & \varphi'(w_{eq})\mathbf{A}_{per} \\ \frac{1}{\tau}\mathbf{I} & -\frac{1}{\tau}\mathbf{I} \end{pmatrix}, \quad \mathbf{A}_{per} = \begin{pmatrix} -1 & 1 & 0 & \cdot & 0 \\ 0 & -1 & 1 & \cdot & \cdot \\ \cdot & \cdot & \cdot & \cdot & 0 \\ 0 & \cdot & \cdot & -1 & 1 \\ 1 & 0 & \cdot & 0 & -1 \end{pmatrix},$$

and this matrix has the same collection of eigenvalues than the matrix associated to the linearized inertial model (expressed by Eq. (10.2)).

Remark 2.13. The stability analysis can be interpreted in terms of modeling. The stability condition reads $\alpha \leq 2$, with $\alpha = 4\varphi'(w_{eq})\tau$. If we introduce $\eta = 1/\varphi'(w_{eq})$, the condition reads

$$\frac{\tau}{\eta} \leq 2.$$

The parameter $\eta = 1/\varphi'(w_{eq})$ quantifies the stiffness of the behavior curve (see Fig. 2.1). A small value reveals a reckless or careless behavior. In the context of car traffic, a driver with a small η typically does not respect safety distances. As expected, it tends to violate the stability condition. While η encodes how the pedestrians (or the drivers) consider themselves, the parameter τ pertains to their actual capabilities. In the context of car traffic, a large τ corresponds to a heavy vehicle with defective brakes, or, if one considers that τ is a reaction time (see model (2.19)–(2.20)), we recover the fact that slow reactions are likely to lead to instabilities.

Chapter 3

Social Force Model, Native and Overdamped Forms

This chapter presents the social force model introduced in the 1990s. Pedestrians are identified with inertial particles, submitted to a forcing term which implements the individual tendencies, and extra forces which account for interactions with other pedestrians (typically the tendency to preserve a certain distance with neighbors). We first present in Section 3.1 the model in its native form (second order in time), and we describe in a second part (Section 3.2) an overdamped version (first order in time). A third section is dedicated to alternative approaches which have been proposed by the Robotics and Computer Graphics communities.

3.1. Inertial Social Force Model

The most popular microscopic model of pedestrian motion was proposed in Helbing *et al.* (1995). We present here the basic form of this model, some extensions which have been proposed, and we address some theoretical issues.

Social force model: assumptions and actual behavior

In its initial version, the model identifies pedestrian with inertial particles submitted to forces. If one denotes by $x_i(t) \in \mathbb{R}^2$ the position of individual i, for $i = 1, \ldots, N$, at time t, and by $u_i = \dot{x}_i$ their velocity, the model is based on Newton's law:

$$m_i \frac{du_i}{dt} = \frac{m_i}{\tau} (U_i - u_i) + \sum_{j \neq i} f_{ij} + \sum_{k} f_{ik}^w, \qquad (3.1)$$

where

- m_i is the mass of individual i;
- U_i is the desired velocity, so that the first term of the right-hand side encodes a force (that corresponds to the motive force of i) which tends to relax to the desired velocity with a characteristic time τ;
- f_{ij} is the force "exerted" by individual j on i (and f_{ik}^w is the force exerted by obstacle k on i) (Fig. 3.1).

The latter notion is the core of the model, and calls for some comments. In case of high congestion, this force may actually correspond to a physical interaction force. Yet, when the distance between individuals (considered as finite size objects) is positive, the effect that is represented by this term results from a complex process: i sees j (possibly anticipating their behavior), and takes the decision to alter their own trajectory, typically to preserve a certain distance with j (see Section 9.2). This overall process, which is far from being mechanical in nature, is encoded in Helbing's model by f_{ij}, which appears in the equation like electrostatic forces would appear for charged particles. For this reason, the overall approach is currently referred to as *Social Force Model*. Notice that, in its native form, the model implements interaction forces which verify the Law of Action–Reaction. Asymmetric interactions can be implemented by multiplying those forcing terms by a correction factor which accounts for the cone of vision associated to each individual (see Eq. (3.4)).

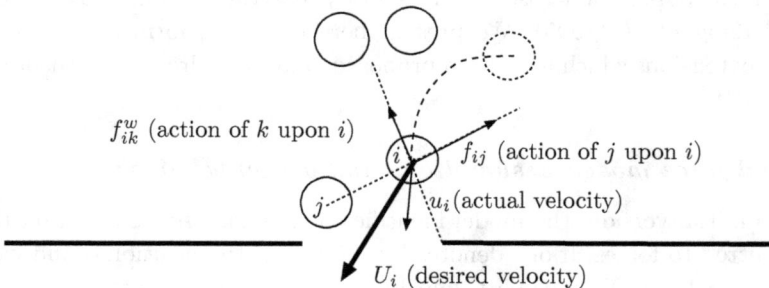

Fig. 3.1. Pedestrians/disks evacuating a room.

Influence of obstacles or walls is accounted for in a similar manner: one may define forces f_{ik}^w to account for the influence of obstacle k upon the motion of individual i.

Isotropic interactions

In the first approach proposed by Helbing, f_{ij} expresses the trend of each person to remain at a certain distance from their neighbors (see Section 9.2). An extra term, also of the repulsive type, was already added in the first version of the model, to handle physical contacts between individuals. The overall repulsion force reads (friction is put off until later)

$$f_{ij} = (-F \exp{(-D_{ij}/\delta)} + \kappa(D_{ij})_-)e_{ij}, \qquad (3.2)$$

with

$$e_{ij} = \frac{x_j - x_i}{|x_j - x_i|}, \quad D_{ij} = |x_j - x_i| - r_i - r_j, \quad F > 0, \quad \delta > 0, \quad \kappa > 0,$$

and $(D_{ij})_-$ is the negative part of D_{ij}, that is D_{ij} whenever $D_{ij} < 0$ (overlapping), and 0 otherwise. The coefficient F (homogeneous to a force) quantifies the social trend of i to keep apart from j. It may actually depend on i (social inner trends), and even on the couple (i, j) (type of the (i, j) relationship), but we shall keep it as F to alleviate notation. The coefficient δ is also involved in qualifying this trend to maintain a certain distance from neighbors. It corresponds to the so-called proxemic distance (see Section 9.2) and it roughly corresponds to the distance below which this term becomes significant compared to its maximal value (when i and j are in contact) (Fig. 3.2).

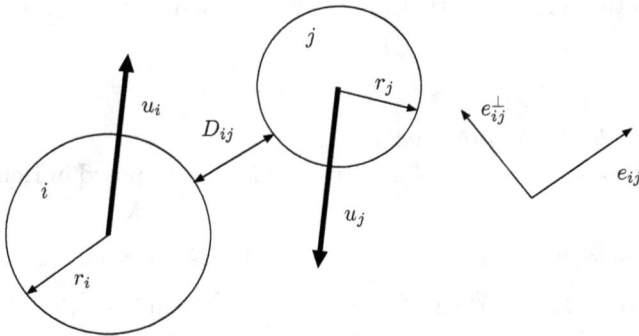

Fig. 3.2.   Notation.

The second term of (3.2) is meant to handle overlapping. It is activated as soon as there is overlapping between the disks ($D_{ij} < 0$), and κ may be seen as a stiffness constant of individuals seen as deformable bodies. Note that both terms contribute to a pull-back force which acts against too deep overlapping.

The effect of obstacles is handled in a similar way.

$$f_{ik}^w = (-F^w \exp\left(-D_{ik}^w/\delta\right) + \kappa \left(D_{ik}^w\right)_-)e_{ik}^w, \tag{3.3}$$

with

$$e_{ik}^w = \frac{x_{ik}^w - x_i}{|x_{ik}^w - x_i|}, \quad D_{ij} = |x_{ik}^w - x_i| - r_i, \quad F^w > 0, \quad \delta^w > 0, \quad \kappa > 0,$$

where x_{ik}^w is the point obstacle k which realizes the distance between obstacle k and x_i.

Remark 3.1. Complying with common usage in the literature, we favored here a representation of obstacles as a set of distinct objects (walls, pieces of furniture, pillars, ...). Yet, it can be more efficient in practice to consider this set as a unique zone which gathers all obstacles. It makes it possible to model the effect of obstacles by means of a unique distance function, that is the distance to the union of obstacles. It may cause some troubles in zones where this distance is not a regular function, but it appears to behave quite efficiently in practice. We refer to Chapter 8 for detailed considerations on those issues.

Friction

An extra term can be added to the interaction force to account for friction. It opposes the relative motion in the transverse direction $e_{ij}^\perp$, and it takes the form

$$f_{ij}^{\text{friction}} = \eta \left(D_{ij}\right)_- \left((u_i - u_j) \cdot e_{ij}^\perp\right) e_{ij}^\perp,$$

where $\eta > 0$ is the friction coefficient.

The following collection of numerical values is proposed in Helbing *et al.* (1995):

$$m = 80 \,\text{kg}, \quad \tau = 0.5 \,\text{s}, \quad F = 2 \times 10^3 \,\text{N}, \quad \delta = 0.08 \,\text{m},$$

$$\kappa = 1.2 \times 10^5 \,\text{kg s}^{-2}, \quad \eta = 2.4 \times 10^5 \,\text{kg m}^{-1}\,\text{s}^{-1}.$$

We also refer the reader to Corbetta *et al.* (2015) for the description of a probabilistic method to estimate the parameter values from experiments.

Mechanical standpoint

Since $-e_{ij}$ is the gradient of D_{ij} with respect to i's position x_i, the first term of the right-hand side of (3.2) can be written

$$-F \exp\left(-D_{ij}/\delta\right) e_{ij} = -\nabla_{x_i}\left(F\delta \exp\left(-D_{ij}/\delta\right)\right).$$

The second term is, in a similar way, the negative gradient of $\kappa\left(D_{ij}\right)_{-}^2 /2$. This remark makes it possible to write the whole system (without friction) in a concise form (ODE in $\mathbb{R}^{2N}$):

$$M\frac{d^2x}{dt^2} = M\frac{du}{dt} = \frac{M}{\tau}\left(U - u\right) - \nabla\left(\Psi^s(x) + \Psi^g(x)\right),$$

where Ψ^s and Ψ^g can be seen as potential energies, the first of which encodes the social tendency to keep at a certain distance from neighbors, whereas Ψ^g corresponds to elastic deformation of individual seen as deformable circular grains. The first term encodes the action of pedestrians, each of which strives to achieve their desired velocity. The mechanical nature of the model is enlightened by the energy balance which can be derived from Newton's principle, by taking the dot product with the velocity itself. Since, by the chain rule,

$$\nabla\Psi^\alpha(x(t)) \cdot \frac{dx}{dt} = \frac{d}{dt}\left(\Psi^\alpha(x(t))\right) \quad \text{for } \alpha = s, \ g,$$

one straightforwardly obtains

$$\frac{d}{dt}\left(\underbrace{\frac{1}{2}Mu \cdot u}_{\text{kinetic energy}} + \underbrace{\Psi^s(x) + \Psi^g(x)}_{\text{potential energies}}\right) = \underbrace{\frac{M}{\tau}\left(U - u\right) \cdot u}_{\text{power of the forcing term}}.$$

It expresses that the total energy, that is the sum of kinetic and potential energies, identifies with the power of forces exerted by the pedestrians to approach their desired paths. Whereas Ψ^g might be considered as a standard elastic energy (stored by deformation of individuals), the "social" energy Ψ^s is some sort of virtual potential which measures the psychological dissatisfaction of individuals forced to accept proximity with neighbors, against their will. Note that, if one disregards the forcing term, the remaining system is of the Hamiltonian type, like a system of inertial charged particles submitted to electrostatic repulsion forces.

Remark 3.2. This standpoint makes it possible to include geometrical features in the model, by defining interaction potentials which are not a function of the distance only. Such an approach is proposed in Helbing and Johansson (2009), where the potential is defined as a function of what the authors call the semi-minor axis of the elliptical potential line. Such a quantity is defined in such a way that it depends on the time step, and the modification vanishes when this time step goes to 0. The corresponding model is thus of the purely discrete type, i.e. it is not the time discretization of a well-identified system of differential equations.

Angular dependence

The purely mechanical character of the native Helbing's model is based on the fact that interaction forces obey the Law of Action–Reaction. This makes clear sense for mechanical interactions between individuals which are in physical contact (the so-called granular forces described previously). It makes less sense for social forces, which result from a cognitive process. A distant individual j is likely to exert some influence on i only if i accounts for j, i.e. sees j. This directionality of influence can be expressed in a smooth way by correcting the social force term. The correcting pre-factor can be defined (see e.g. Helbing and Johansson, 2009) as

$$w(\theta_{ij}) = \left(\lambda_i + (1 - \lambda_i) \frac{1 + \cos(\theta_{ij})}{2} \right), \qquad (3.4)$$

where θ_{ij} is the angle at which i (assumed to look ahead along his desired velocity) sees j (see Fig. 3.3):

$$\theta_{ij} = \arccos\left(\frac{U_i}{|U_i|} \cdot e_{ij} \right).$$

The parameter $\lambda_i \in [0, 1]$ quantifies this directional dependence (with $\lambda_i = 1$ the fully isotropic case is recovered).

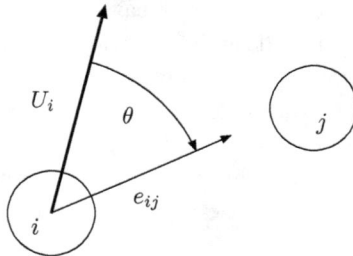

Fig. 3.3.   Angle of vision.

This correction obviously rules out the Law of Action–Reaction. Thus, it deeply changes the mechanical nature of the model, transforming particle-like pedestrian to active entities who take decisions based on their perception.

Social force model: mathematical issues

Existence and uniqueness of a solution

We consider here the social force model (3.1) in its native form:

$$m_i \ddot{x}_i = \frac{m_i}{\tau}(U_i - \dot{x}_i) + \sum_{j \neq i} f_{ij}\left(+\sum_k f_{ik}^w\right), \quad i = 1, \ldots, N.$$

We shall disregard momentarily the last term, which implements interactions with obstacles, in order to alleviate notation. We furthermore assume that f_{ij}, which encodes interactions between individuals i and j, is a function of their distance $D_{ij} = |x_j - x_i| - 2r$ (where r is the common radius of our circular individuals). We write the system in position/velocity variables, to obtain the first order system

$$\frac{dx_i}{dt} = u_i, \quad i = 1, \ldots, N,$$

$$m_i \frac{du_i}{dt} = \frac{m_i}{\tau}(U_i - u_i) + \sum_{j \neq i} f_{ij}(D_{ij}), \quad i = 1, \ldots, N. \tag{3.5}$$

Let us first consider the case where f_{ij} is given by (3.2), i.e.

$$f_{ij} = (-F \exp(-D_{ij}/\delta) + k(D_{ij})_-)e_{ij},$$

with $e_{ij} = (x_j - x_i)/|x_j - x_i|$. The fact that this expression involves the negative part of D_{ij} is not an issue, since $D \mapsto (D)_-$ is a Lipschitzian function. Cauchy–Lipschitz theory therefore ensures the existence and uniqueness of a maximal solution defined on some time interval $[0, T^\star)$, as soon as initial conditions are chosen "reasonably". The only difficulty comes from the fact that $(x_i, x_j) \longmapsto |x_j - x_i|$ is not a smooth function on the "diagonal" $x_i = x_j$.

Proposition 3.1. *Assume an initial condition (x, u) is given in $\mathbb{R}^{2N} \times \mathbb{R}^{2N}$, such that $x_i \neq x_j$ for $i \neq j$. The Cauchy problem (3.5) admits a unique maximal solution $t \mapsto (x(t), u(t)) \in \mathcal{U} \times \mathbb{R}^{2N}$, where*

$$\mathcal{U} = \{x = (x_1, \ldots, x_N) \in \mathbb{R}^{2N}, \ i \neq j \Longrightarrow x_i \neq x_j\}.$$

Proof. This is direct consequence of Cauchy–Lipschitz theorem (Theorem A.1), in the space-time set $\mathcal{U} \times \mathbb{R}^{2N} \times [0, +\infty)$. Indeed, all functions of the distances are Lipschitzian, and the distances are locally smooth in the open set $\mathcal{U}$. $\square$

It is worth noticing that the maximal solution given by the previous theorem may be defined only locally because, as already mentioned, the distance function $(x_i, x_j) \mapsto |x_j - x_i|$ stops being smooth when $x_j = x_i$. Therefore it cannot be excluded that the solution only exists as far as centers are all distincts.

Remark 3.3. The non-existence of a global solution does not prevent from actually doing long-lasting computations. A numerical problem may occasionally happen whenever two positions x_i and x_j are undistinguishable (i.e. equal up to machine precision). In the latter situation, division by $|x_j - x_i| = 0$ is likely to stop the program. Since this critical situation has little chance to happen in practice, it is to be expected that computations will continue beyond the theoretical solution lifetime. Yet, the computed solution may make little sense in terms of modeling. A typical situation is represented in Fig. 3.4. On the left, the two disks are almost superimposed, and e_{12} points rightward. On the right, which may correspond to the computed configuration at the next time step, e_{12} points leftward, so that interaction forces are suddenly reversed. The solution tells that 1 and 2 have gone *through each other*, and the crossing is accompanied by an abrupt change in interaction forces.

Increasing the coefficient in the repulsion term k (second term in the right-hand side of (3.2)) is likely to make less likely the occurrence of such critical situations, but it is difficult to be sure *a priori* that they will not happen. Consider for instance the situation presented in Fig. 3.8. Increasing the number of people in the line will tend to crush the crowd against the

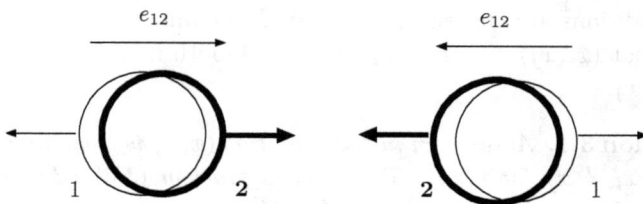

Fig. 3.4.   Non-smoothness of $(x_1, x_2) \mapsto e_{12}$.

wall, up to the situation presented above. This very case of people heading toward a wall, or a closed door, is of course caricatural, but similar situations may happen in more realistic situations, e.g. with people heading toward a jammed exit. Another attempt to have a proper global solution may consist in smoothing the interaction force in the neighborhood of the critical zone. This approach may lead to a rigorous well-posedness result, but does not make any sense from the modeling standpoint: it would amount to consider that repulsion forces vanish when people collapse.

The only way to obtain a satisfying result from both mathematical and modeling standpoint consists in implementing the fact that repulsion forces become arbitrary large when the distance approaches some prescribed minimal value. We shall express such a condition in terms of potential energy. Let us set at $d_{\min} > 0$ a minimal value for $|x_j - x_i|$ (in other words D_{ij} is forced to take values larger than $d_{\min} - 2r < 0$). The crucial assumption we make is the following: interaction forces derive from a potential which blows up to infinity when $|x_j - x_i|$ approaches $d_{\min}$:

$$f_{ij} = -\nabla_{x_i} V(|x_j - x_i|) = V'(|x_j - x_i|)\, e_{ij}, \qquad (3.6)$$

where V maps $(d_{\min}, +\infty)$ to $\mathbb{R}^+$, V is regular (in $C^2(d_{\min}, +\infty)$), and

$$\lim_{d \to d_{\min}^+} V(d) = +\infty. \qquad (3.7)$$

This blow-up assumption will make it possible to keep the solution away from the critical zone, thereby ruling out the possibility that the solution may cease to exist. Those consideration are properly expressed in the next proposition.

Proposition 3.2. *Let (x^0, u^0) be given in $\mathcal{U}_{d_{\min}} \times \mathbb{R}^{2N}$, where is $\mathcal{U}$ is defined by*

$$\mathcal{U}_{d_{\min}} = \{x = (x_1, \ldots, x_N) \in \mathbb{R}^{2N},\ i \neq j \Longrightarrow |x_j - x_i| > d_{\min} > 0\}.$$

Assume furthermore that interaction forces meet assumptions (3.6) and (3.7). The Cauchy problem (3.5) with initial data (x^0, u^0) admits a unique maximal solution $t \mapsto (x(t), u(t)) \in \mathcal{U}_{d_{\min}} \times \mathbb{R}^{2N}$, and this solution is global, i.e. defined over $[0, +\infty)$.

Proof. Existence and uniqueness of a maximal solution is given again by Theorem A.1. To prove that this solution is global, we shall establish that it is bounded in positions and velocities, and that positions remain away from the boundary of $\mathcal{U}_{d_{\min}}$. We denote by $t \mapsto (x(t), u(t))$ the maximal solution, defined on the time interval $[0, T^\star)$ with $T^\star$ possibly finite. We aim at establishing that, if $T^\star$ is finite, (x, u) remains bounded in $[0, T^\star)$, and that x remains at a certain distance from the boundary of $\mathcal{U}_{d_{\min}}$ bounded from below. By Proposition A.1 it will ensure that $T^\star = +\infty$, i.e. the solution is global.

The core of the proof relies on energy arguments. We take masses equal to 1 to alleviate notation. The momentum equations read

$$\frac{du_i}{dt} = \frac{1}{\tau}(U_i - u_i) + \sum_{j \neq i} V'(|x_j - x_i|)\, e_{ij}.$$

We multiply each of those by the corresponding u_i, and we sum over all individuals:

$$\sum_{i=1}^{N} \frac{du_i}{dt} \cdot u_i + \frac{1}{\tau}\sum_{i=1}^{N} |u_i|^2 = \frac{1}{\tau}\sum_{i=1}^{N} U_i \cdot u_i - \sum_{i<j} V'(|x_j - x_i|)\, e_{ij} \cdot (\dot{x}_j - \dot{x}_i).$$

Noticing that $e_{ij} \cdot (\dot{x}_j - \dot{x}_i) = d\,|x_j - x_i|\,/dt$, we may write

$$V'(|x_j - x_i|)\, e_{ij} \cdot (\dot{x}_j - \dot{x}_i) = V'(|x_j - x_i|)\frac{d\,|x_j - x_i|}{dt} = \frac{d}{dt} V(|x_j - x_i|).$$

We thus have

$$\frac{d}{dt}\underbrace{\left(\sum_{i=1}^{N}\frac{1}{2}|u_i|^2 + \sum_{i<j} V(|x_j - x_i|)\right)}_{E(t)} + \frac{1}{\tau}\sum_{i=1}^{N}|u_i|^2 = \frac{1}{\tau}\sum_{i=1}^{N} U_i \cdot u_i,$$

where $E(t)$ is the total energy at time t (sum of kinetic and interaction potential energies). Now using $U_i \cdot u_i \leq (|U_i|^2 + |u_i|^2)/2$, we get, by integrating the previous equation over $(0, t)$ (with $t < T^\star$),

$$E(t) - E(0) + \frac{1}{2\tau}\int_0^t |u_i|^2 \leq \frac{1}{2\tau}\int_0^t |U_i|^2,$$

from which we deduce that, *if $T^\star$ is finite*, the total energy is bounded over $[0, T^\star)$ (the previous inequality holds for any $t < T^\star$, and the right-hand side is bounded). This energy is the sum of two non-negative terms, both of which are therefore non-negative as well. The kinetic energy is therefore bounded, so that u is bounded, and so is x, by integration of u over $[0, T^\star)$. Furthermore, the interaction potential energy is bounded, and so is each of its constitutive terms $V(|x_j - x_i|)$. Thanks to the blow-up assumption on

V we deduce that distances $|x_j - x_i|$ are bounded from below by a quantity $d_{\min} + \eta$, with $\eta > 0$ (the contrary would rule out boundedness of potential energies). We established that the maximal solution remains in a compact of $\mathcal{U}_{d_{\min}} \times \mathbb{R}^{2N}$ over the time interval $[0, T^\star)$, which is (by Theorem A.1) in contradiction with the finite character of $T^\star$. As a consequence, $T^\star = +\infty$, which means that the solution is global. □

Spurious oscillations

The social force model accounts for inertia: individuals are considered as massive particles submitted to forces, some of which depend on their positions. Such a system is likely to produce spurious oscillations. We propose here to investigate the possibility that such oscillations may appear. The typical situation is that of two persons facing each other, and heading in opposite directions (see Fig. 3.5, left). We shall actually consider the equivalent situation of a single individual heading leftward toward a rigid wall (Fig. 3.5, right), the action of which is encoded by a repulsion force of the type (3.2). We denote by x the position of the disk center, r its radius, and $-U < 0$ the desired velocity. In this oversimplified situation, the model writes

$$m\ddot{x} = \frac{m}{\tau}(-U - \dot{x}) + F\exp\left(-\frac{x - r}{\delta}\right), \qquad (3.8)$$

or, written in $(x, u = \dot{x})$ variables,

$$\dot{x} = u,$$
$$\dot{u} = \frac{1}{\tau}(-U - u) + \frac{F}{m}\exp\left(-\frac{x - r}{\delta}\right). \qquad (3.9)$$

Let us first establish that, as expected, all trajectories converge to the unique equilibrium point x_{eq}, that is the only position which balances the leftward tendency and the wall rightward repulsion.

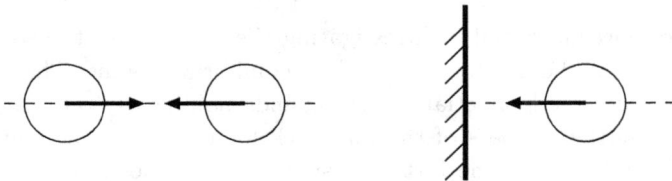

Fig. 3.5.   Conflicting situations.

Proposition 3.3. *For any initial data x^0, $u^0 = \dot{x}(0)$, the system* (3.9) *admits a unique maximal solution, that is global (i.e. defined over $[0, +\infty)$). This solution converges toward the unique equilibrium point $(x_{\text{eq}}, 0)$, where x_{eq} is the unique solution to*

$$F \exp\left(-\frac{x_{\text{eq}} - r}{\delta}\right) = \frac{m}{\tau} U. \tag{3.10}$$

Proof. Let us first note that, since $x \mapsto \exp(-(x - r)/\delta)$ is decreasing, and maps $\mathbb{R}$ on $(0, +\infty)$, Eq. (3.10) indeed defines a unique equilibrium point x_{eq}. The system verifies the assumptions of Cauchy–Lipschitz Theorem (Theorem A.1), thus it admits a unique maximal solution. Now consider the energy functional

$$\Phi \; : \; (x, u) \longmapsto \frac{u^2}{2} + \varphi(x),$$

where φ is such that

$$\varphi'(x) = \frac{U}{\tau} - \frac{F}{m} \exp\left(-\frac{x - r}{\delta}\right).$$

This function of x is strictly convex, and it admits a unique minimizer at x_{eq}, so that $(x_{\text{eq}}, 0)$ is the unique minimizer of Φ over $\mathbb{R}^2$. It holds that, for any solution $t \mapsto (x(t), u(t))$,

$$\frac{d}{dt} \Phi(x(t), u(t)) = \nabla\Phi \cdot \begin{pmatrix} \dot{x} \\ \dot{u} \end{pmatrix} = u\varphi' + u\left(-\frac{1}{\tau} u - \varphi'\right) = -\frac{u^2}{\tau} \leq 0.$$

Therefore Φ is a Lyapunov function for the system over $\mathbb{R}^2$ (see Definition A.4). Furthermore, the value of Φ is strictly decreasing along any solution that is not the constant one $x(t) \equiv x_{\text{eq}}$, it is therefore a strict Lyapunov function, from which we deduce convergence of the trajectory toward $(x_{\text{eq}}, 0)$ (thanks to Proposition A.3, page 179). $\qquad\square$

We are now interested in investigating the behavior of the solution to this equation in the neighborhood of its equilibrium point. The difficulty is that the force (exponential term) depends nonlinearly on the position itself, so that the stiffness of the pull-back force towards the equilibrium point. We shall make some further assumptions to alleviate notation. We consider the situation where the repulsion forces exactly balance the selfish tendency when the disk is in contact with the wall, i.e. $F = mU/\tau$.

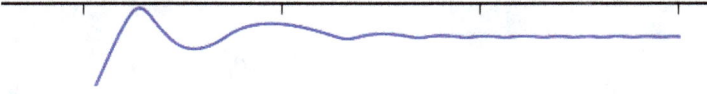

Fig. 3.6. Spurious oscillations.

The equilibrium position is then $x_{\text{eq}} = r$. The linearized system at $(x_{\text{eq}}, 0)$ is then characterized by the matrix

$$\mathbf{A} = \begin{pmatrix} 0 & 1 \\ -\dfrac{U}{\delta\tau} & -\dfrac{1}{\tau} \end{pmatrix}, \quad \text{with characteristic polynomial } \lambda^2 + \frac{1}{\tau}\lambda + \frac{U}{\delta\tau} = 0.$$

The solution will therefore present damped *oscillations* as soon as

$$\frac{1}{\tau^2} - 4\frac{U}{\delta\tau} < 0 \iff \frac{4U\tau}{\delta} > 1.$$

This behavior is illustrated by Fig. 3.6: a single individual tends to head to the top toward a fixed wall, and the position with respect to the wall is represented as the time goes on (from left to right). Oscillations are clearly observable in this situation.

Social force model: critical discussion

The social force model presented in the previous section is undoubtedly the most popular microscopic model, and it is the core of many commercial software. As previously described, the model is known to lead in some situation to unwanted phenomena, like unrealistic oscillations of individuals (see e.g. Kretz, 2015). Those artefacts can actually be suppressed by considering a *simpler* version of the model, where inertia is neglected. It is presented in Section 3.2. We refer to Cristiani and Sahu (2016), Section 4.2, for a discussion on first- and second-order ODE models.

3.2. Overdamped Social Force Model

We present here a non-inertial (overdamped) version of the social force model. It can be formally obtained from (3.1) by considering interaction forces are scaled like $1/\tau$, and by having τ go to 0. We obtain a set of Ordinary Differential Equations of the first order in time with respect to the positions. We denote by $x_i = x_i(t) \in \mathbb{R}^2$ the position of i at time t and by U_i their desired velocity. It reads

$$u_i = \frac{dx_i}{dt} = U_i + \sum_{j \neq i} W_{ij}, \qquad (3.11)$$

where the W_{ij} are no longer forces, but rather *corrections* of the desired velocities: W_{ij} is the correction of i's velocity induced by individual j. Like in the inertial model, φ_{ij} is meant to implement the tendency of individuals to stay apart, a natural definition is therefore

$$W_{ij} = -U \exp\left(-D_{ij}/\delta\right) e_{ij},$$

where U quantifies the effect of this interaction, and δ plays the role of the interpersonal distance (see Section 9.2). Both may depend on i, and possibly on both i and j, since they implement a social reaction of i with respect to j. Like in the inertial case, an extra repulsive term can be added to account for physical interactions, i.e. whenever $D_{ij} = |x_j - x_i| - r_i - r_j$ becomes negative.

Gradient flow framework

Under some assumptions detailed below, the model above takes the very particular form of a gradient flow, i.e. it can be written

$$\frac{dx}{dt} = -\nabla \Psi(x),$$

where Ψ maps $\mathbb{R}^{2N}$ to $\mathbb{R}$. From a modeling standpoint, it amounts to say that the crowd can be considered as a single point in a high-dimensional space ($\mathbb{R}^{2N}$), which flows along the steepest slope path on the landscape defined by Ψ.

This calls for strong assumptions on desired velocities and interactions:

- The desired velocity field derives from a potential,

$$U_i = -\nabla V_i(x_i).$$

- Full interdependence: each individual is potentially influenced by all others (including those that are behind him).
- For any couple (i, j), the interaction term derives from a potential which depends on the distance between i and j only:

$$W_{ij} = \nabla_{x_i} V_{ij}(D_{ij}).$$

If the previous assumptions are verified, the evolution problem (3.11) takes the form

$$\frac{dx}{dt} = -\nabla\Psi(x) = -\nabla\left(\sum_i V_i(x_i) + \sum_{i \neq j} V_{ij}(D_{ij})\right). \qquad (3.12)$$

The global dissatisfaction function is the sum of a term which encodes individual tendencies (each individual tends to lower their own dissatisfaction V_i) and a second term pertaining to interactions (two individuals are unsatisfied when they are too close to each other). For instance, in the case of an evacuation, one may consider that all individuals have the same dissatisfaction function, that is proportional to the distance D to the exit, and the desired velocity is then defined accordingly (see Chapter 8). Since the gradient of the potential has to be homogeneous to a velocity, we shall define this potential as the product of the typical speed of a free individual times the distance to the exit, i.e.

$$V_i(x_i) = U\, D(x_i).$$

As for interactions, it is natural to define the mutual dissatisfaction in such a way that its gradient vanishes when the distance goes to infinity, and significantly affects the individual tendency when the distance becomes smaller than the interpersonal distance δ (see Section 9.2). One possible choice is then

$$V_{ij}(D_{ij}) = V(D_{ij}) = \kappa U\, \delta\, \exp\left(-D_{ij}/\delta\right), \qquad (3.13)$$

where $\kappa > 0$ (typically of the order of 1) is a dimensionless prefactor, the role of which is to tune up the effect of this term.

The instantaneous evolution results from a trade-off between individual tendencies and inter-individual competition, in particular in the case where their respective potentials V_i tend to drive them toward the same target.

A toy example of gradient flow evolution

In order to illustrate the gradient flow framework, and more generally the effect of the interaction force (3.13) in ODE models, we consider a very simple conflicting situation, with two individuals 1 and 2 moving on the real line, each of which tries to reach the origin (one from the left, one from the right), while being reluctant to get too close to each other.

Denoting by x_1 and x_2 the positions, and by $U > 0$ the free speed, we define

$$V(x) = U\,|x|\,, \quad V_1(x_1) = V(x_1), \quad V_2(x_2) = V(x_2),$$
$$V_{12} = \delta U \exp\left(-D_{12}/\delta\right).$$

Now assume that x_1 and x_2 are initially located at $-M$ and M, respectively. By symmetry, we shall have $x_1 = -x_2$ for all times, and we get for x_1 the differential equation

$$\dot{x}_1 = U\left(1 - \exp\left(\frac{2(x_1 + r)}{\delta}\right)\right).$$

If $M \gg \delta$, individual 1 will progress from $x_1(0) = -M$ rightward at a roughly constant speed U. When $|x_1|$ becomes of the order of δ, x_1 slows down, and his velocity continues to decrease to 0 while he approaches the equilibrium point $x_1^\infty = -r$, which corresponds to the case of two individuals in contact. This contact is a trade-off between individual tendencies which tend to push each of them in the direction of the other, and social tendencies which tend to keep them apart (Fig. 3.7).

If one wants to reinforce the repulsive trend, in order to obtain a equilibrium distance closer to the interpersonal distance, the interaction potential has to be multiplied by a dimensionless prefactor $\kappa > 1$. In this case, the equilibrium is attained when

$$x_1 = -r - \frac{\delta}{2}\log \kappa.$$

As a consequence, for $\kappa = e$, the distance $D_{12} = |x_2 - x_1| - 2r = 2\,|x_1| - 2r$ at equilibrium is exactly the comfort distance δ.

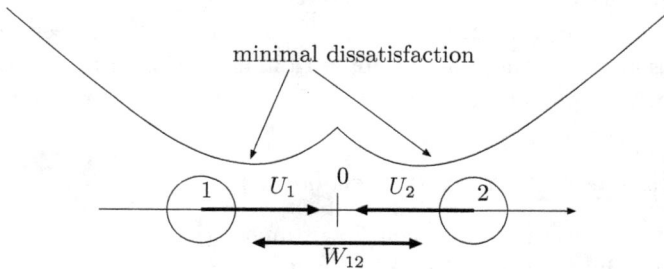

Fig. 3.7.   Conflict of the gradient flow type.

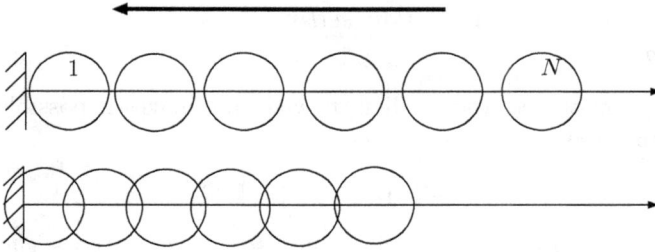

Fig. 3.8.   Heading straight into the wall (soft setting).

It is important to notice than the typical distance ensured by this inter-action model is highly dependent on the global situation, in particular on the number of individuals which are involved in direct interactions. To illustrate this remark, let us consider the situation represented in Fig. 3.8: N individuals aim at moving leftward at velocity U, toward a closed door at 0. The evolution can be put in a gradient flow framework by consid-ering that each individual i has a dissatisfaction $U|x|$, and interacts with neighbor j (for $j = i - 1$ and $j = i + 1$) through the interaction potential

$$V_{ij} = \kappa \delta U \exp\left(-D_{ij}/\delta\right).$$

The equilibrium point (which corresponds to the minimum of the global dissatisfaction functional) verifies an equation of the force balance type. Interpreting accordingly velocities as forces, it is obvious that the rightward reaction of the wall upon 1 exactly counterbalances all the leftward efforts exerted by individuals. As a straightforward consequence, for any given pre-factor κ, individual 1 will turn out to enter the wall when the number N goes to infinity. The value of x_1 may even become negative, which makes little sense from a modeling standpoint.

Remark 3.4. If one aims at favoring a better control of distances between individuals, in a robust way with respect to the size of clusters, one may favor a stiffer expression of the interaction potential, like

$$V_{ij}(D_{ij}) = V(D_{ij}) = \kappa U \exp\left(-\frac{D_{ij} - \delta}{\varepsilon}\right), \tag{3.14}$$

where ε is a stiffness parameter. When ε goes to 0, the action of this term vanishes for $D_{ij} > 0$, and goes to infinity otherwise, which will force the distance to remain of the order of δ or larger.

Illustration of the gradient flow setting for a single individual

As already mentioned, the gradient flow setting makes it possible to comprehend the crowd as a single point

$$x = (x_1, \ldots, x_N) \in \mathbb{R}^{2N},$$

which slides along the steepest slope line (see Eq. (3.12)) associated to its global dissatisfaction

$$\Psi(x) = \sum_i V_i(x_i) + \sum_{i \neq j} V_{ij}(D_{ij}),$$

where the V_i (typically proportional to the distance to the exit for an evacuation scenario) encode individual trends, and the V_{ij}'s (defined for example by (3.13)) quantify the mutual dissatisfaction of being too close to each other. The high dimension makes it difficult to fully picture out this crowd as a "hyper water drop" sliding along the steepest descent of the landscape function Ψ. We propose here to illustrate how this setting formalizes the conflict between individual trends and social tendencies in a simpler setting. We consider the situation where a single individual moves in an environment of obstacles (or "frozen" individuals) which are treated by adding repulsive potentials. In this spirit, we define the individual dissatisfaction as $V(x) = U|x|$, where the position vector x is computed with respect to an origin located at the center of the exit door (see Fig. 3.9). Obstacles (treated as frozen pedestrians) are disks of radius r centered at $y_1, \ldots, y_p$. The interaction potentials are defined by Eq. (3.13), so that the

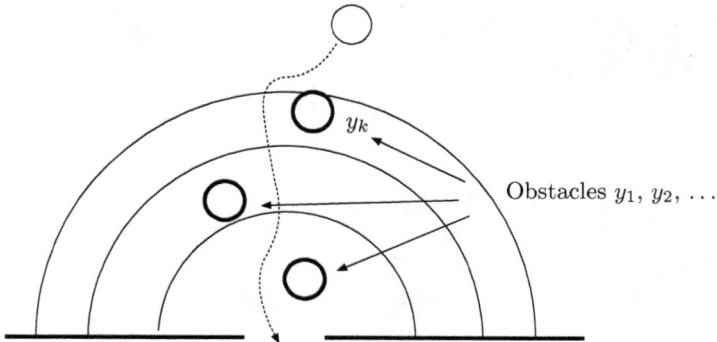

Fig. 3.9.   Gradient flow evolution for a single individual.

full dissatisfaction function is

$$\Psi(x) = U\,|x| + \kappa\,U\,\delta \sum_{k=1}^{p} \exp\left(-D_{1k}/\delta\right), \quad D_{1k} = |y_k - x_1| - 2r.$$

Figure 3.10 represents isolines of the corresponding potential for different numbers of obstacles (0, 1, 2, and 6), with the parameters

$$U = 1, \quad \kappa = 2, \quad \delta = 0.8\,\text{m}, \quad r = 0.4\,\text{m},$$

with an individual potential defined as the distance to the center of an exit door of width 1 m.

Angular dependence

Model (3.11) is written in its native form, under the assumption that an individual is likely to be influenced by all the others. Like in the previous section, angular dependence can be accounted for by multiplying the correction term by a pre-factor (defined for instance by (3.4)), or more drastically by restricting the sum to indices j corresponding to individuals that influence i. One may for instance define an angle θ, a length ℓ, the associated cone of vision $C = C(x_i, U_i, \theta, \ell)$ (see Fig. 3.11), and restrict the sum to I_i, that is the set of all those indices j such that x_j lies in C.

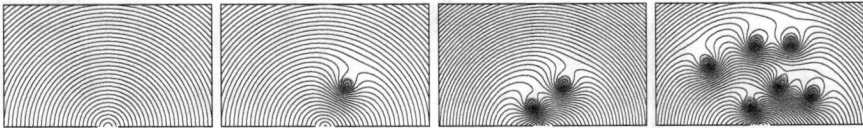

Fig. 3.10.   Potential isolines.

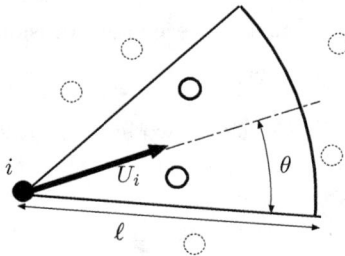

Fig. 3.11.   Cone of vision.

Critical discussion on the gradient flow setting

The gradient flow approach gives a very particular structure to the evolution problem, which makes it much more tractable from the mathematical standpoint (well posedness, equilibrium points, long term evolution). Yet, it relies on strong assumptions, some of which are questionable from the modeling standpoint. In particular, the gradient character of interactions automatically ensures symmetry of interactions (overdamped version of the Law of Action–Reaction). Thus, it rules out the possibility to account for the fact that some interactions might be strictly one-way (1 sees 2, while 2 does not see 1).

Another limitation pertains to desired velocity fields. While the gradient flow setting is natively adapted to evacuation situations, or more generally to situation when each pedestrian aims at approaching a given zone, it is ruled out in other cases. Consider for example the situation of Pilgrims at Mecca, who are required to circumambulate seven times around the Kaaba. The associated velocity field can obviously not be written as the gradient of a scalar potential.

3.3. Alternative Approaches

We end this chapter by mentioning alternative strategies to develop two-dimensional microscopic models for crowd motion. Those approaches do not take the form of systems of Ordinary Differential Equations, since they are natively discrete in time, and as such they do not fully fit in the scope of this book. Yet, they present some similarities with the previous approaches, or at least time discretization of the previous approaches, since they consist in setting rules to define instantaneous velocities which account for individual tendencies (goal seeking), and interaction with nearby agents (collision avoidance). Besides, they prove to be very efficient in practice, and lead to highly realistic scenarios (Computer Graphics standpoint), and they are also very robust in terms of collision avoidance (Robotics standpoint).

The starting point is the notion of Obstacle Velocity, introduced in Fiorini and Shiller (1998). Consider two agents i and j (identified to disks of some radius r) at some instant, and set the velocity of j to be equal to u_j. Considering that j evolves in the future at constant velocity u_j, the induced Velocity Obstacle is defined as the set $\mathrm{VO}_{i|j}(v_j)$ of all those velocities u_i such that, if i chooses and maintains u_i in the future, a collision will happen at some time in the future. It leads to a way to update velocities by handling agents one after the other, picking for each agent a new velocity which is outside of all the VO's associated to nearby agents. In order to circumvent the oscillations induced by the previous approach, the so-called Reciprocal Velocity Obstacle (RVO) is proposed in Van den Berg *et al.* (2008) to simulate real-time multi-agent navigation. It can be interpreted as some sort of relaxation/stabilization of the previous procedure. In the case of two agents, the velocity u_i of agent i is updated by picking a new velocity $\tilde{u}_i$ which is the average between u_i and a velocity outside $\text{VO}_{i|j}(v_j)$. The multi-agent case is simply handled by extending the relaxation strategy to the set of agents, i.e. by dealing with agents one after the other.

More recently, the so-called Optimal Reciprocal Collision Avoidance (ORCA) approach has been introduced in Van den Berg *et al.* (2011). Consider two nearby agents i and j and a time $\tau > 0$. The basis of the approach consists in defining a set W of admissible couples for the velocities u_i and u_j, i.e. velocities which do not lead to a collision during time τ. Setting a set V_i of velocities for i induces constraints on the velocity for j. More precisely, requiring that $(u_i, u_j) \in W$ for any $u_i \in V_i$ imposes that u_i belongs to some set $V_j = C_{i|j}^\tau(V_i)$. Such sets V_i and V_j are called reciprocally collision-avoiding and reciprocally maximal if $V_j = C_{i|j}^\tau(V_i)$ and $V_i = C_{j|i}^\tau(V_j)$. The idea consists in building such sets V_i and V_j in an optimal and "fair" way, by requiring that the intersections of V_i and V_j with any ball centered at the desired velocity U_i and U_j (respectively) have the same measure (fairness), and that this measure is in some way the largest among all possible reciprocally collision-avoiding and maximal pairs (optimality). The evolution algorithm is then based on a relaxation approach, which can be interpreted as an *explicit* scheme: agents are handled one after the other, and their velocities are built in order to respect the previous principles (considering at each time that the velocities of others are given). We refer to Van den Berg *et al.* (2011) for further details on the formulation and implementation aspects.

Note that this approach presents some similarities with the granular models presented in Chapter 4. Those models can be interpreted as an instantaneous instance of the ORCA approach. In particular, they are all based on globally defining a set of admissible velocities (with constraints on the relative velocities between nearby agents). Yet, in opposition with the granular models, the actual evolution depends on the order in which agents are considered, so that inferring a time-continuous deterministic evolution process under this procedure does not seem possible, but this approach is unquestionably relevant, and in some way richer than standard ODE approaches, in that it explicitly accounts for the ability of pedestrians to *anticipate* the evolution of their neighborhood, rather than passively reacting to the present observations.

Chapter 4

Granular Models

This chapter is dedicated to crowd motion models of the *granular* type: each individual is identified to a hard disk of a prescribed size, subject to a non-overlapping constraint with their neighbors. The approach relies on a desired velocity for each individual (the velocity they would take if they were alone), and the global velocity field shall be defined as the closest to the desired one among all those *feasible* fields (fields which do not lead to over-lapping of disks). Section 4.1 presents a one-dimensional toy model, which is extended in Section 4.2 to the more realistic two-dimensional setting. A numerical strategy is proposed in Section 4.3, together with preliminary tests. Section 4.5 addresses mathematical issues raised by this approach. Those issues call for sophisticated theoretical tools, borrowed from Convex Analysis and the theory of Differential Inclusions, but a full understanding of those mathematical aspects is not mandatory to understand the model itself and its numerical simulation. The chapter ends by a discussion (Section 4.6).

4.1. One-Dimensional Model

We consider N individuals subject to move on a straight line. Individuals are identified to rigid segments of common length $2r > 0$ (see Fig. 4.1), the centers of which are denoted by

$$q_1 < q_2 < \cdots < q_N.$$

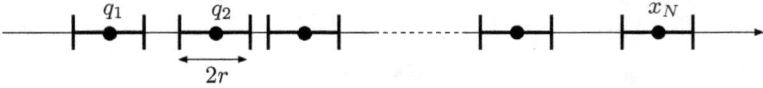

Fig. 4.1.   One-dimensional setting.

The non-overlapping assumption constrains $q = (q_1, \ldots, q_N)$ to belong to the set of feasible configurations

$$K = \{q = (q_1, \ldots, q_N) \in \mathbb{R}^N, \ q_{n+1} - q_n \geq 2r, \ \forall n = 1, \ldots, N - 1\}.$$

We consider that each individual tends to follow a desired velocity U_i (possibly varying in time), so that $U = (U_1, \ldots, U_N)$ is the global desired velocity. We aim at preserving the non-overlapping constraint, which amounts to prescribe a non-negative relative velocity for segments in contact. We accordingly define the set of feasible velocities as

$$C_q = \{v = (v_1, \ldots, v_N) \in \mathbb{R}^N, \ q_{n+1} - q_n = 2r \implies v_{n+1} - v_n \geq 0\}.$$

Model 4.1. The one-dimensional granular model reads

$$\frac{dq}{dt} = P_{C_q}(U),$$

where P_{C_q} stands for the Euclidean projection on the closed convex cone C_q.

Saddle-point formulation

Let us consider a fully congested situation: each individual is in contact with his neighbors, i.e. $q_2 - q_1 = 2r$, $q_3 - q_2 = 2r, \ldots$.

Projecting U on C_q amounts to minimize

$$J(v) = \frac{1}{2} \sum |v_i - U_i|^2$$

over the set C_q of feasible velocities, which can be written

$$C_q = \{v = (v_1, \ldots, v_N) \in \mathbb{R}^N, \ Bv \leq 0\},$$

where B is the $(N - 1) \times N$ matrix

$$B = \begin{pmatrix} 1 & -1 & 0 & \cdot & \cdot & 0 \\ 0 & 1 & -1 & \cdot & \cdot & \cdot \\ \cdot & \cdot & \cdot & \cdot & \cdot & 0 \\ \cdot & \cdot & & \cdot & 1 & -1 \end{pmatrix}. \tag{4.1}$$

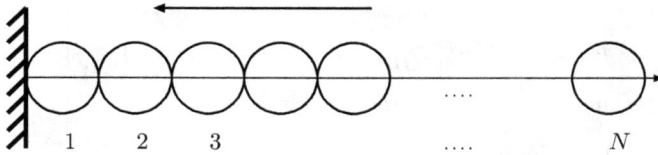

Fig. 4.2. Heading straight into the wall (hard setting).

This problem fits in the assumptions of Proposition A.4: there exists $p \in \mathbb{R}_+^{N-1}$ such that

$$\begin{cases} \nabla J(u) + B^\star p = 0, \\ Bu \leq 0. \end{cases}$$

In other words, there is a collection of non-negative pressures $p_{n,n+1}$, for $n = 1, \ldots, N - 1$, such that, for any two persons n and $n + 1$ in contact

$$u_n = U_n + p_{n-1,n} - p_{n,n+1}.$$

The Lagrange multipliers can therefore be interpreted as interaction forces between people in contact, and the actual velocity of individual n is their desired velocity corrected by the net action of his neighbors, $p_{n-1,n} - p_{n,n+1}$.

Let us consider the following extreme, yet illustrative, situation: N individuals in a row tend to go leftward with the same velocity $-U < 0$, with a wall located at 0 (see Fig. 4.2). The solution is obviously static: all actual velocities are 0, which is made possible by positive pressures $p_{0,1}$, $p_{1,2}, \ldots, p_{N-1,N}$, where p_{01} corresponds to the action of the wall on individual 1. A straightforward computation exhibits a hydrostatic like pressure field:

$$p_{N-1,N} = U, \quad p_{N-2,N-1} = 2U, \ldots, \quad p_{0,1} = NU.$$

This example illustrates an important feature of highly congested crowds: the action of individuals sum up to create high interaction forces. In the present case, individual 1 is crushed with an intensity that grows to $+\infty$ with the number of people behind him.

4.2. Two-Dimensional Model

We represent individuals as rigid disks of common radius r. The position vector is

$$q = (q_1, q_2, \ldots, q_N) \in \mathbb{R}^{2N}.$$

The set of feasible configurations (no overlapping) is defined as

$$K = \{q \in \mathbb{R}^{2N}, D_{ij} = |q_j - q_i| - 2r \geq 0, \forall i \neq j\}. \tag{4.2}$$

We consider a collection of desired velocities

$$U = (U_1, \ldots, U_N).$$

In the simplest setting (asocial and interchangeable individuals), each U_i depends on the position of i only. In the latter case we have that $U_i = U_0(q_i)$, where U_0 is the desired velocity fields which is shared by all individuals. More complex models can be elaborated by writing $U = U(q)$, which expresses that the desired velocity of an individual depends upon their position, but also upon the position of others.

The cone of feasible velocities associated to a configuration q is then

$$C_q = \{v, D_{ij}(q) = 0 \Rightarrow e_{ij} \cdot (v_j - v_i) \geq 0\}$$
$$= \{v, D_{ij}(q) = 0 \Rightarrow G_{ij} \cdot v \geq 0\}, \tag{4.3}$$

where

$$G_{ij} = \nabla D_{ij}(q) = (0, \ldots, 0, -e_{ij}, 0, \ldots, 0, e_{ij}, 0, \ldots, 0) \in \mathbb{R}^{2N} \tag{4.4}$$

is the gradient of the distance from i to j. The vector e_{ij} is the center-to-center unit vector (see Fig. 4.3)

$$e_{ij} = \frac{q_j - q_i}{|q_j - q_i|}.$$

Note that $G_{ij} \in \mathbb{R}^{2N}$ has only four non-zero components, which correspond to the degrees of freedom of individuals i and j. The model expresses the fact that the effective velocity is the closest to the desired one among all feasible velocities, i.e.

$$\frac{dq}{dt} = P_{C_q} U(q), \tag{4.5}$$

where P_{C_q} is the Euclidean projection on C_q, defined by (4.3).

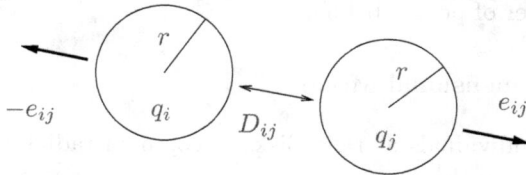

Fig. 4.3.   Notation.

Saddle-point formulation

The projection problem (4.5) consists in minimizing

$$J(v) = \frac{1}{2} |v - U|^2, \tag{4.6}$$

of the set C_q of feasible velocities. The constraint can be written in matrix form:

$$C_q = \left\{ v \in \mathbb{R}^{2N}, \ Bv \leq 0 \right\},$$

where each row of B expresses a non-overlapping constraint between two disks which are in contact in the current configuration: it identifies with $-G_{ij}$, where $G_{ij} = \nabla D_{ij}$ is defined by (4.4). We shall denote by N_C the number of contacts, i.e. the number of couples (i, j), with $i < j$, such that $D_{ij} = 0$. In this setting, B belongs to $\mathcal{M}_{N_C, 2N}(\mathbb{R})$.

The minimization problem can be written in a saddle-point form, which takes the form of a discrete Darcy problem:

Proposition 4.2. *The minimization of J (defined by (4.6)) over C_q (defined by (4.3)) can be expressed in a saddle-point form:*

$$\begin{cases} u + B^\star p = U, \\[2mm] Bu \leq 0, \\[2mm] p \geq 0, \\[2mm] p \cdot Bu = 0. \end{cases} \tag{4.7}$$

where $p \geq 0$ is meant element-wise, i.e. $p_{ij} \geq 0$ for all i and j in contact. More precisely, if u minimizes J over C_q, there exists $p \in \mathbb{R}^{N_C}$ such that (4.7) is verified. Conversely, if $(u, p) \in \mathbb{R}^{2N} \times \mathbb{R}^{N_C}$ solves (4.7), then u minimizes J over C_q.

Proof. This is a direct consequence of Proposition A.4. $\qquad\qquad\square$

Note that the last equation of (4.7) expresses that

$$-\sum_{i \sim j} \underbrace{G_{ij} \cdot u}_{\geq 0} \ \underbrace{p_{ij}}_{\geq 0} = 0.$$

As a consequence, each term $G_{ij} \cdot u \, p_{ij}$ is equal to zero, and System (4.7) can be written is a more detailed way:

$$\begin{cases} u_i - \sum_{i \sim j} p_{ij} G_{ij} = U_i, \\[2mm] -G_{ij} \cdot u \leq 0 \quad \forall i \sim j, \\[2mm] p \geq 0, \\[2mm] G_{ij} \cdot u > 0 \Longrightarrow p_{ij} = 0, \end{cases} \tag{4.8}$$

where $i \sim j$ means that i and j are in contact ($D_{ij} = 0$).

Remark 4.1. While the minimizer u in the previous proposition is uniquely defined, the pressure p is not necessarily unique. Some configurations (like the one represented in Fig. 4.10) are such that there are more contacts than degrees of freedom, i.e. B has more rows (N_C) than columns ($2N$). The matrix $B^\star$ is then not injective, and the Lagrange multiplier is not unique. To be precise, the sole fact that $B^\star$ is not injective is not equivalent to non-uniqueness, because p is required to be non-negative. Yet, it suffices to consider a compressed situation such that a fully positive solution exists, i.e. all p_{ij} are positive. Consider now a non-trivial $w \in \ker B^\star$. Then for ε sufficiently small, $p + \varepsilon w$ is also a solution.

4.3.   Numerical Scheme

We are interested here in approximating solution to (4.5). The scheme we propose is based on a very simple idea, which was introduced in the context of granular flows (Maury, 2006). Consider an admissible configuration $q \in K \subset \mathbb{R}^{2N}$ (where K is defined by (4.2)), and a time step $\tau > 0$. Applying to this configuration a velocity field v during τ yields the configuration $q + \tau v$. The idea consists in replacing the admissibility of the new configuration, which reads

$$D_{ij}(q + \tau v) \geq 0 \quad \forall i, j, \ i \neq j,$$

by first-order expansions of those non-overlapping constraints, i.e.

$$D_{ij}(q) + \tau \nabla D_{ij}(q) \cdot v \geq 0 \quad \forall i, j, \ i \neq j. \tag{4.9}$$

The time-stepping scheme follows from this simple approximation. Considering a time step $\tau > 0$ and an initial configuration q^0, successive approximations are built as follows. Assume that q^k is known (approximation at time step k). The next configuration is computed according to

$$u^{k+1} = \arg\min_{C^\tau_{q^k}} \frac{1}{2}|v - U|^2, \qquad (4.10)$$

where $C^\tau_{q^k}$ is the "discretized" set of feasible velocities

$$C^\tau_{q^k} = \left\{ v \in \mathbb{R}^{2N}, \ D_{ij}(q^k) + \tau \nabla D_{ij}(q^k) \cdot v \geq 0 \quad \forall i \neq j \right\}. \qquad (4.11)$$

Let $i \sim j$ express the fact that the constraint between i and j is likely to be active.[1] The constraints (4.9) at the current configuration q^k can be expressed as follows:

$$\forall i \sim j, \quad -\nabla D_{ij}(q^k) \cdot v \leq \frac{D_{ij}(q^k)}{\tau} \iff Bv \leq z,$$

where the vector z contains the $D_{ij}(q^k)/\tau$, and B is a matrix with $2N$ columns, and a number of rows which is equal to the number of potential active contacts. Each row of B corresponds to a constraint between i and j, and it contains the $2N$-vector $-\nabla D_{ij}(q^k)$.

The saddle-point formulation writes

$$\begin{cases} u + B^\star p = U, \\ Bu \leq z, \\ p \geq 0, \\ p \cdot Bu = 0. \end{cases} \qquad (4.12)$$

The minimization problem can be solved by a Uzawa algorithm (see (A.4)). This algorithm (the index ℓ corresponds to the inner iterations of the optimization algorithm, whereas k corresponds to time indices) writes

$$p^{\ell+1} = \Pi_+(p^\ell + \rho(B(U - B^\star p^\ell) - z)), \qquad (4.13)$$

and the velocities $u^\ell = U - B^\star p^\ell$ can be shown to converge to the minimizer of $|v - U|^2/2$ over $C^\tau_{q^k}$ (defined by (4.11)), see Appendix A.2.

Remark 4.2. In this setting, the condition (A.5) on the parameter ρ can be checked to be fairly independent of the time step τ, and on the number of individuals. Indeed, the matrix A is the identity matrix (i.e. $\alpha = 1$

---

[1]In practice, numerical efficiency requires to restrict the constraints to couples which are likely to saturate the constraints, e.g. by disregarding individuals which are far away (i.e. more than a few diameters) from each other. Suppressing those constraints does not change the outcome of the minimization problem, while improving its efficiency.

in (A.5)). Besides, the norm of B is bounded by the max over the rows of the sum of the absolute values of the entries, which scales like unity (the entries represent unit vectors in $\mathbb{R}^2$) times the number of potential active constraints. If the constraints are written only for disks which are close to each other, the number of neighbors of a given disk is uniformly bounded.

Remark 4.3. If one considers that each optimization step is computed exactly, the overall time discretization process described above can be proven to converge to the solution to the evolution problem (properly defined in Section 4.5), as detailed in Venel (2011).

Remark 4.4 (Stopping criterion). In practice, the Uzawa algorithm is run (i.e. iterations (4.13) are performed) until some criterion is met. It is here natural to prescribe some tolerance in the inequality constraint $Bu^\ell - z \leq 0$, i.e. to stop the algorithm as soon as all elements of this vector are smaller than a positive threshold value. Tuning up this threshold in a robust way with respect to all parameters (time step, number and size of individual, desired velocities, ...) is a delicate issue. A reasonable choice can be made by noticing that the positive part of $\tau(Bu^\ell - z)$ quantifies the overlapping between disks. If we denote by d the typical diameter of those disks, a natural criterium is then

$$\tau(Bu^\ell - z) = \tau(B(U - B^\star p^\ell) - z) \leq \varepsilon d,$$

where the inequality is meant element-wise, and ε is a tuneable dimensionless parameter which quantifies the amount of overlapping that is tolerated, relatively to the size of the disks.

4.4. Numerical Experiments

As detailed in Section 9.8, the presence of an obstacle upstream a narrow exit may improve the evacuation process. The present granular model makes it possible to reproduce this effect in some situations, as illustrated by Fig. 4.4. Snapshots of an evacuation are presented, both in an empty room (left) and in a room with a triangular obstacle close to the door. In the reference simulation, a static jam appears. The obstacle seems to prevent its creation. Let us add that this effect is very sensitive to parameters and geometrical characteristics. In particular, an obstacle slightly farther away from the exit has no apparent effect on the evacuation, whereas placing it too close to the exit increases the probability of jam creation. We refer to Sections 4.6, 9.8 and 10.2 for additional considerations on those issues.

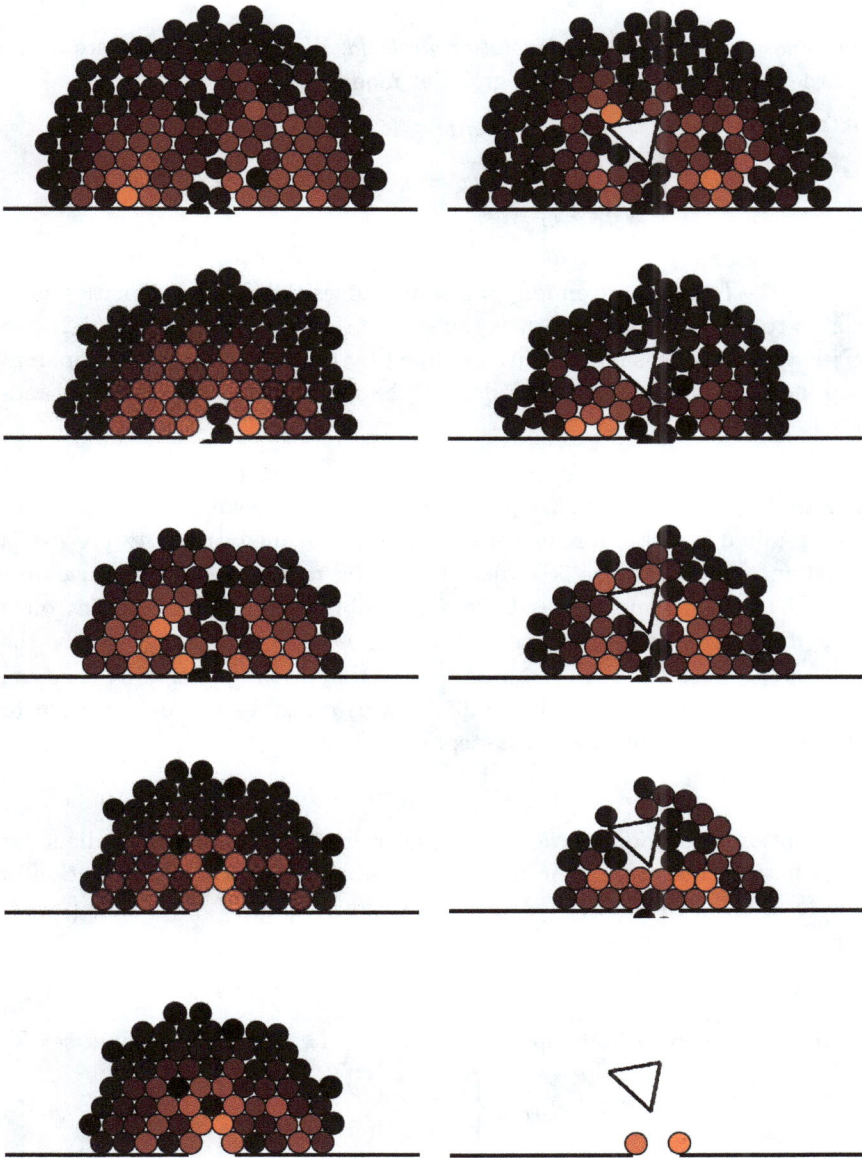

Fig. 4.4.   Fluidizing role of an obstacle.

4.5. Mathematical Issues

Well posedness of the time evolution problem. We are interested here in the mathematical structure of the granular model (4.5):

$$\text{Find } t \longmapsto q(t) \in K \text{ such that}$$

$$q(0) = q_0 \in K \text{ (given)}$$

$$\frac{dq}{dt} = P_{C_q} U(q) \text{in } [0, T],$$

where $q \mapsto U(q)$ is a given mapping which gives the desired velocities associated to any configuration, K is the set of feasible configurations (4.2) and C_q is the set of feasible velocities defined by (4.3). In spite of its apparent simplicity, this problem raises delicate issues from the theoretical standpoint.

Moreau's framework. The approach proposed in Moreau (1977) makes it possible to give a sound framework to the crowd evolution problem. In this seminal paper, Moreau considered the so-called *sweeping process*, a point in a Hilbert space is subject to remain within a moving convex set $t \mapsto K(t)$, while moving as little as possible. The approach relies on a time discretization process, called *catching up algorithm*. Let $\tau > 0$ be the time step. We denote by q^k an approximation of the position at time $k\tau$. Successive approximations are built by projecting the current position to the convex set at the next time step:

$$q^{k+1} = P_{K((k+1)\tau)}(q^k).$$

The notion of *subdifferential*, which generalizes the notion of gradient for non-smooth convex function makes it possible to express the projection in the form of an inclusion. For any convex function $\Psi : H \mapsto \mathbb{R} \cup \{+\infty\}$, one defines

$$\partial\Psi(q) = \{v, \Psi(q) + v \cdot h \leq \Psi(q+h), \forall h\}. \tag{4.14}$$

In the case where Ψ is the indicatrix function I_K of a closed convex set K:

$$I_K(q) = \begin{cases} 0 & \text{if } q \in K, \\ +\infty & \text{if } q \notin K, \end{cases} \tag{4.15}$$

it holds that (it is a straightforward consequence of the characterization of the projection on a closed convex set)

$$\partial I_K(q) = \{q' - q, \ q = P_K(q')\}.$$

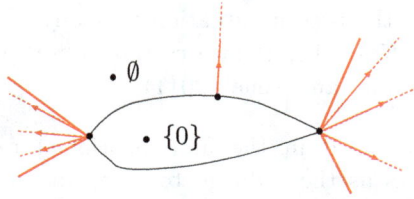

Fig. 4.5.   Outward normal cone to a convex set.

Figure 4.5 represents this outward normal cone in various situations: for interior points it reduces to $\{0\}$; it is empty for points which lie outside of K; for any point on the boundary of K, it corresponds to the outward normal half line in the smooth case, and to a non-trivial close convex cone for angular extremal points. The fact that q^{k+1} is the projection of q^k on K can be expressed

$$\frac{q^{k+1} - q^k}{\tau} \in -\partial I_{K((k+1)\tau)}(q^{k+1}).$$

The latter can be interpreted as the (implicit) time discretization of the inclusion

$$\frac{dq}{dt} \in -\partial I_{K(t)}(q).$$

This approach can be extended to more general situations, in particular when the point q is animated by a spontaneous velocity U. The corresponding catching-up algorithm (in the case of a fixed convex set) can be written

$$\begin{cases} \tilde{q}^{k+1} = q^k + \tau U^k, \\ q^{k+1} = P_K(\tilde{q}^{k+1}). \end{cases} \tag{4.16}$$

The approach is not directly applicable to crowd motion, because K is not a convex set.[2]

The constructive algorithm obviously necessitates that the projection is well-defined in a neighborhood of K only, and it therefore applies to sets for which this projection is well-defined. Such sets are called *prox-regular* (see e.g. Thibault (2003)).

---

[2]Unless one considers the case of a single individual in a convex room with no exit, which is of little interest.

One can check in the present situation that any point q of $\mathbb{R}^{2N}$ which lies a distance from K smaller than a certain $\eta > 0$ projects in a unique manner on K (see Maury and Venel, 2011).

Remark 4.5. As detailed in the aforementioned reference, the prox-regularity degenerates as the radii go to zero, and when the number of individuals goes to $+\infty$. It means that the projection on K is properly defined on a neighborhood of K, the width of which goes to zero with the size of individuals.

If one assumes that the desired velocity is bounded, the algorithm (4.16) is therefore well-posed for τ sufficiently small, and one can show (Maury and Venel, 2011) that the sequence of discrete trajectories converge to a solution of the differential inclusion

$$\frac{dq}{dt} \in -\partial I_K(q) + U \quad \text{for a.e. } t, \tag{4.17}$$

to the price of an extension of the notion of subdifferential to indicatrix function of non-convex sets. The right notion is that of *Fréchet subdifferential*, which is a relaxed version of (4.14):

$$\partial \Psi(q) = \{ w \in H, \ \Psi(q) + h \cdot w \leq \Psi(q+h) + o(h), \forall h \}. \tag{4.18}$$

Alternative approach, connection to maximal monotone operators. We propose here a more academic reformulation of the evolution problem, based on the notion of *polar cone*. The polar cone to C_q, which we denote N_q, is defined as

$$N_q = (C_q)^\circ = \{ w \in \mathbb{R}^{2N}, \ w \cdot v \leq 0, \forall v \in C_q \}.$$

We now use a classical result in convex analysis (Moreau, 1962), that is the decomposition of the identity in a Hilbert space as the sum of projection operators on two mutually polar cones (see Fig. 4.6):

$$\mathbf{I} = P_{C_q} + P_{N_q}.$$

We therefore have $U = P_{C_q} U + P_{N_q} U$ (see Fig. 4.6), so that the differential equation implies $dq/dt - U \in N_q$, which we write

$$\frac{dq}{dt} + N_q \ni U.$$

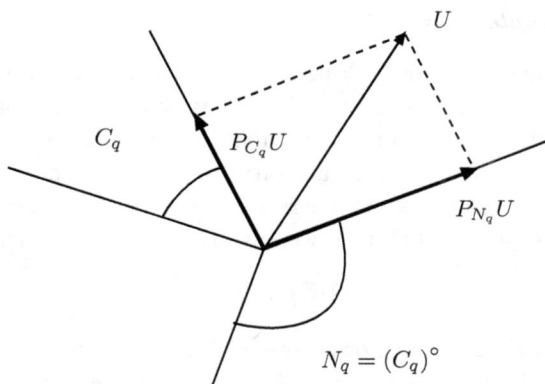

Fig. 4.6.   Polar cones and Moreau's decomposition.

Remark 4.6. Farkas' lemma gives an explicit expression of N_q, that is

$$N_q = \left\{ -\sum p_{ij} G_{ij}, \ \ p_{ij} \geq 0, \ \ D_{ij}(q) > 0 \implies p_{ij} = 0 \right\},$$

which is consistent with the saddle-point formulation (4.8).

The last step in reformulating the evolution equation consists in using the notion of Fréchet subdifferential, defined by (4.18). Consider the so-called *indicatrix* function I_K of the set of feasible configurations, defined by (4.15).

It can be established that the outward normal cone N_q identifies with the Fréchet subdifferential of the indicatrix function of K at q, so that the evolution problem can be written

$$\frac{dq}{dt} + \partial I_K(q) \ni U(q). \tag{4.19}$$

This type of equation has been intensively studied in the 1970s, in the case where K is *convex* (see Brezis, 1973). When K is convex, the subdifferential ∂I_K is known to be *maximal monotone*, which essentially means that

$$\forall (x_1, x_2) \in H \times H, \ \ \forall (y_1, y_2) \in \partial I_K(x_1) \times \partial I_K(x_2), \ \ (y_2 - y_1) \cdot (x_2 - x_1) \geq 0.$$

As a consequence, it can be established that $(I_d + \partial I_K)^{-1}$ is a single-valued operator, and it is a contraction. This makes it possible to prove existence and uniqueness of a solution to (4.19) as soon as $U(\cdot)$ has Lipschitz regularity.

Underlying laplace operator

Consider a saturated queue of N pedestrians. The constraint matrix is given by (4.1). This matrix expresses a discrete version of $-\partial_x$ (opposite of the divergence operator in dimension 1), and $B^\star$ corresponds to ∂_x (gradient). In the case where all constraints are saturated (if one supposes for instance that the desired velocities are decreasing: the persons ahead wish to go slower than the persons behind), we have that $Bu = 0$, which implies

$$BB^\star p = BU.$$

The square matrix $BB^\star$, the order of which is $N-1$, is exactly the matrix of the discrete Laplace operator with Dirichlet boundary conditions. The pressure field therefore turns out to be the solution to a discrete *Poisson* problem, with a source term. The source term quantifies the spontaneous tendency of individuals to violate the non-overlapping constraint.

In the two-dimensional setting, the analogy is more delicate. Consider $q \in K$ (see Fig. 4.7), and the associated matrix B, each row of which expresses a constraint of the type

$$-G_{ij} \cdot v \leq 0,$$

where G_{ij} is the gradient of $D_{ij} = |q_j - q_i| - r_i - r_j$ with respect to $q = (q_1, \ldots, q_N)$. Consider a collection p of dual variables. The mapping

$$p \mapsto -B^\star p$$

realizes the action of those interaction forces on the primal network, where the native degrees of freedom (positions of disk centers) are defined.

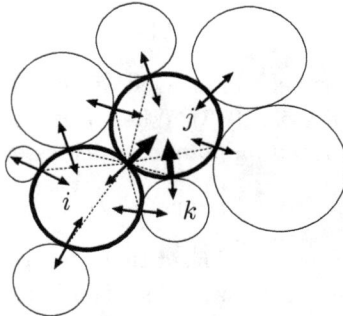

Fig. 4.7.   Unstructured stencil.

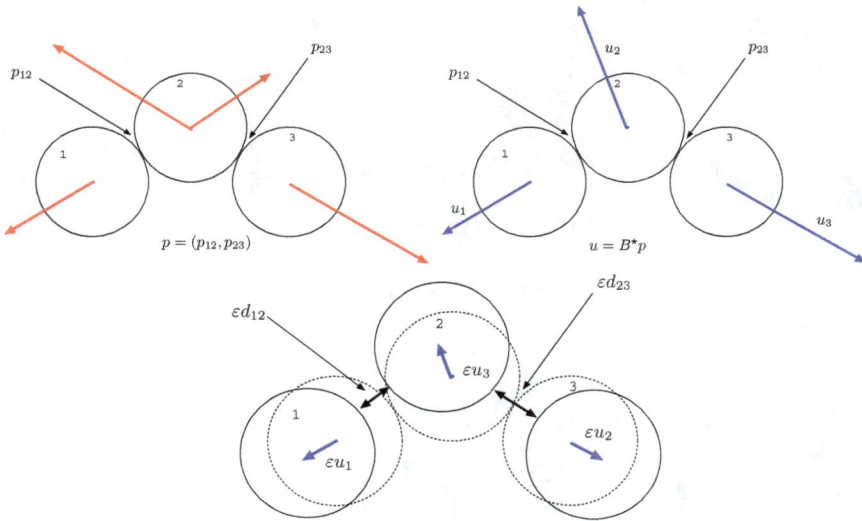

Fig. 4.8.   Geometric representation of $p \longmapsto -B^\star p$.

Figure 4.8 represents this mapping in the case of three grains and two contacts. In the case of a structured situation (for instance a cartesian network, or a triangular one like in Fig. 4.10), a uniform pressure field has a zero discrete gradient on interior points.[3] Yet, in the general case (when the local granular arrangement does not present any symmetry), this property is ruled out. For example in the situation represented in Fig. 4.7, one may straightforwardly check that the sum of unit vectors pointing inward each of the two grains in contact is not zero. The bi-dimensional case presents another feature. Consider the cluster represented on Fig. 4.10. The number of disks is 14, thus the number of degrees of freedom is 28, and the number of active contacts (dimension of the dual space) is 29. As a consequence, the kernel of $B^\star \in \mathcal{M}_{29,28}(\mathbb{R})$ is non-trivial: there exists non-zero pressure fields which induce a zero resultant force on the grains. A consequence of this pathological behavior is that the discrete operator, although it is the discrete analogous to a Laplace operator defined on the dual network (represented on the right of Fig. 4.9), does *not* verify the maximum principle. In other words, there may exist pressure fields p such

---

[3]One recovers the discrete version of the fact that a constant function has a zero gradient.

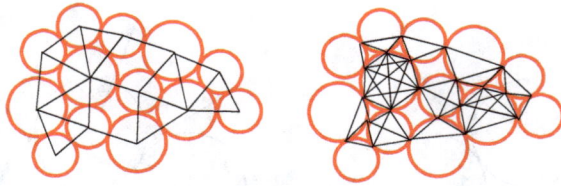

Fig. 4.9.   Primal (left) and dual (right) networks.

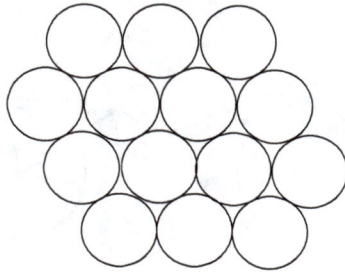

Fig. 4.10.   Hyperstatic configuration: 28 degrees of freedom vs. 29 constraints.

that $BB^\star \geq 0$ (i.e. the pressure tends to increase all distances), whereas some individual pressures are negative (attractive forces between some individuals).

Gradient flow framework

Under some assumptions on the desired velocity field, problem (4.5) takes the form of a gradient flow. In other words, the configuration evolves along the steepest descent direction with respect to some potential. In all generality, this structure stems from the following assumption: the desired velocity field

$$U = (U_1, U_2, \ldots, U_N)$$

is the (opposite of the) gradient of a function $x \mapsto \Phi(x)$, i.e.

$$U(q) = -\nabla \Phi(q), \tag{4.20}$$

or, equivalently,

$$U_i(x) = -\nabla_{q_i} \Phi(q) \quad \forall i = 1, \ldots, N.$$

This condition is met in particular when each desired velocity U_i is the gradient of some common potential at q_i. Since this "potential" can be interpreted as a *dissatisfaction* that agent i tends to lower, it will be usually identified with the distance to some common objective (like the exit of a room), and we shall therefore denote it by $D(\cdot)$. The assumption therefore writes $U_i = -\nabla D(q_i)$. In this case, the global potential/dissatisfaction is simply defined as the sum of individual dissatisfactions, i.e.

$$\Phi(q) = \sum_{i=1}^{N} D(q_i), \quad U_i = -\nabla_{q_i}\Phi = -\nabla_{q_i}D(q_i).$$

Since it is the sum of functions depending on individual positions, the latter potential does not account for interactions between agents. Hard-sphere interactions can be implemented by introducing the full dissatisfaction function

$$\Psi(q) = \sum_{i=1}^{N} D(q_i) + I_K(q),$$

where I_K (see (4.15)) is the indicatrix function of the set K of feasible configurations (defined by (4.2)). In this setting, the evolution model (4.5) can be written in a gradient form, using the notion of Fréchet subdifferential (4.18):

$$\frac{dq}{dt} \in -\partial\Psi(q) = -\partial\left(\sum_{i=1}^{N} D(q_i) + I_K(q)\right).$$

The model can therefore be seen as a steepest descent evolution, with respect to a global dissatisfaction function which is the sum of two terms:

(1) Individual tendencies: the term $\Phi(x)$ is the sum of individual dissatisfactions;
(2) Hard sphere interactions $I_K(x)$: the indicatrix term triggers an "infinite dissatisfaction" of the crowd as soon as two disks overlap. This term *de facto* prevents overlapping.

Remark 4.7. More sophisticated tendencies, including collective ones, may be implemented by modifying the behavioral dissatisfaction function. For instance, one can account for the tendency of two individuals i and j to remain close to each other by adding an extra term, depending of q_i and q_j, which increases when D_{ij} increases (both i and j are dissatisfied when their distance is large).

4.6.   Critical Discussion

Individual behavior

The model presented in this chapter is very crude in terms of modeling human behavior. The virtual individuals tend to follow a purely selfish strategy, disregarding other pedestrians. In this native form, it does make sense in very particular situations only, e.g. in panic situations so extreme that self-preservation predominates over all other tendencies. Let us remark, though, that the desired velocity field can be defined in such a way that more social (and possibly *cooperative*) tendencies are accounted for. From this standpoint, the granular approach can be seen as a simple ingredient to account for physical collisions and mechanical interaction between individual, and this ingredient can be combined with more sophisticated models of human behavior, through the definition of the desired velocity.

Symmetric interaction

The granular model relies on a projection of the desired velocity on the set of admissible velocities, in a least square sense. It induces a symmetric character of interactions. This native symmetry is highlighted by the saddle-point formulation of the constrained minimization. Consider two individuals i and j in contact. Equation (4.8) restricted to those two guys reads

$$u_i = U_i - p_{ij}\, e_{ij}, \quad u_j = U_j + p_{ij}\, e_{ij},$$

which has a strong *Law of Action and Reaction* flavor: p_{ij} quantifies the modification of desired velocities for both individuals, in opposite directions. This symmetry can be distorted by attributing different masses to individuals. The "mass" in this non-inertial context quantifies the will, or strength, of the individuals. From a mathematical standpoint, it simply consists in using a weighted ℓ^2 to define the projection, i.e. use

$$|u|_M = \sqrt{\sum_{i=1}^{N} m_i\, |u_i|^2},$$

rather than the standard Euclidean norm, where

$$M = \mathrm{diag}(m_1, m_1, m_2, m_2, \ldots, m_N, m_N) \in \mathcal{M}_{2N}(\mathbb{R})$$

is the mass matrix. It makes the modifications of desired velocities depend on the relative mass of two individuals implicated in a contact, but it does not really make the interaction asymmetric, since the notion of interaction force remains relevant. In particular, when two individuals are in contact, both are affected by this contact. It reveals the mechanical nature of the granular model, which is meant to account for *physical* contacts between individual. As a consequence, this model does not make sense to reproduce the motion of a crowd where individuals remain at a certain distance from each other.

Faster-is-Slower effect: a mathematical standpoint

We investigate here the possibility to *prove* in some way that the microscopic granular model is able to recover the so-called Faster-is-Slower (FiS) effect. We consider an evacuation process in this framework, with a desired velocity field U. The actual instantaneous velocity u is the primal part of the solution to the saddle-point problem (4.7). Now consider a highly congested situation like the one represented in Fig. 4.11. We propose to formulate the FiS issue in the following terms: consider a person (i.e. a disk) of index i on the front of the cluster, i.e. close to the exit. We denote by n_i the direction of their desired velocity. Now consider that any other individual j has the ability to slightly change their speed, i.e. by changing their desired velocity to $\beta_j U_j$, where β_j is a coefficient close to 1. The FiS issue amounts to investigate whether taking β_j *larger* than 1 for some individuals might *decrease* the velocity of i in the desired direction, thereby decreasing the flow near the exit. To formalize this question, we first consider the actual velocity field associated to the desired velocity field U at some instant, i.e. u solves system (4.7). We eliminate all those rows of B that do not correspond to active constraints. We keep the same notation for the reduced B and the associated reduced vector p of Lagrange multipliers.

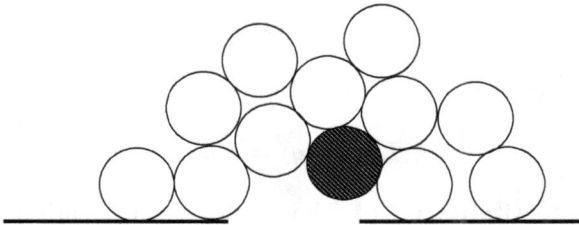

Fig. 4.11.   Congested evacuation.

The system writes

$$\begin{cases} u + B^\star p = U, \\ Bu = 0, \end{cases} \qquad (4.21)$$

where all pressures are positive. The velocity can be eliminated, which yields

$$BB^\star p = BU,$$

which is the discrete counterpart of a standard Poisson problem $(-\Delta p = g)$.

We now consider the velocity as a function u_β of $\beta = (\beta_1, \ldots, \beta_N)$, the vector of speed coefficients, i.e. the desired velocity of individual j is $\beta_j U_j$, where β_j varies around 1. We denote by $\beta \odot U$ the vectors of perturbed desired velocities. Admitting that the network of active contacts remains the same for small variations of β, it holds that

$$BB^\star p_\beta = B(\beta \odot U), \quad u_\beta = \beta \odot U - B^\star p_\beta.$$

The objective function we are interested in is

$$u_\beta \cdot n_i = \beta_i |U_i| - B^\star p_\beta \cdot n_i, \quad n_i = (0, \ldots, 0, U_i/|U_i|, 0, \ldots, 0). \qquad (4.22)$$

We shall disregard variations of speed for the person i, so that the first term above is constant, and we therefore define the objective function as $J(\beta) = -B^\star p_\beta \cdot n_i$.

The core of the approach relies on a co-called *Lagrangian*, that is a function of the state variable p, the control variable β, and a dual variable q. It is defined as the sum of the objective function (velocity of i along the desired direction) expressed in the state variable p (uncoupled from the control variable β), and a weak expression of the state equation $BB^\star p_\beta = B(\beta \odot U)$:

$$L(p, \beta, q) = -B^\star p \cdot n_i + (BB^\star p - B(\beta \odot U)) \cdot q.$$

As a consequence, for any β and any q,

$$L(p_\beta, \beta, q) = J(\beta).$$

We differentiate with respect to β both sides of this identity:

$$D_\beta J = D_p L \circ D_\beta p_\beta + D_\beta L. \qquad (4.23)$$

The approach consists in choosing an adjoint variable q such that $D_p L = 0$, which circumvents the difficulty to explicitly estimate $D_\beta p_\beta$. Let $\tilde{p}$ denote a variation in the variable p. It holds that

$$D_p L \, \tilde{p} = -B^\star \tilde{p} \cdot n_i + (BB^\star \tilde{p}) \cdot q = (-Bn_i + BB^\star q) \cdot \tilde{p}.$$

The adjoint problem is designed in order to vanish the previous expression, it therefore reads

$$BB^\star q = Bn_i. \qquad (4.24)$$

On the other hand, it holds that

$$D_\beta L \, \tilde{\beta} = B(\tilde{\beta} \odot U) \cdot q = -(U \odot B^\star q) \cdot \tilde{\beta},$$

where $U \odot B^\star q$ stands here for an N-dimensional vector, the entries of which are the $U_j \cdot (B^\star q)_j$'s. We finally obtain

$$\nabla J(\beta) = -U \odot B^\star q, \quad i.e. \quad \frac{\partial J}{\partial \beta_j} = U_j \cdot (B^\star q)_j,$$

where q is the solution to the adjoint problem (4.24). This expression of the gradient makes it possible to investigate the FiS effect. Indeed, if an entry of $\nabla J(\beta)$ is negative, i.e. if there exists an index j such that $\partial J / \partial \beta_j < 0$, it means that the corresponding individual j will *reduce* the speed of i in the desired direction by *increasing* their own desired velocity, i.e. by increasing β_j. It can be checked that, in highly congested situations, some coefficients are indeed negative, as illustrated by Fig. 4.12. Each snapshot has been realized as follows: the grain in black correspond to the person which is focused on (person i in the previous notation system). The objective function is the actual speed of i in the desired direction, see Eq. (4.22). For each agent $j \neq i$, we compute the derivative of the actual velocity of person i with respect to speed variations of j. The red color corresponds to positive values: each red grain is such that, if the person tends to accelerate their pace, it will induce an increase of the speed of i. More interesting are the blue grains: each of them corresponds to a person who has a counter-intuitive influence upon i. More precisely, the blue corresponds to *negative* values of the gradient, which means that, if a blue person accelerates, it will induce a reduction of ith speed.

   To complement this representation, we have also computed, for each agent j, the gradient of the actual velocity of person i with respect to velocity variations of j (not only the speed as previously, changes of direction

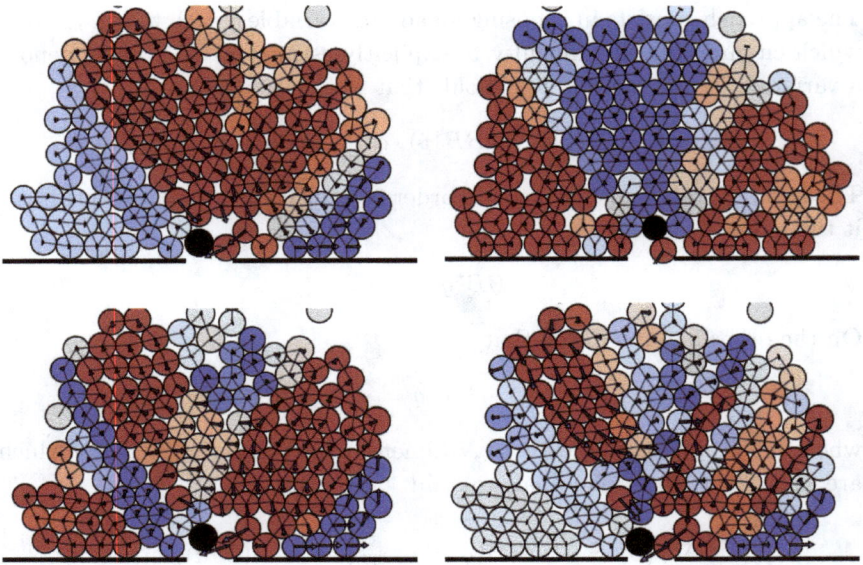

Fig. 4.12.   FiS effect.

are also considered). The black arrow at the center of each agent represents this gradient. It means that if agent j modifies their velocity along this direction, it *increases* the objective function, i.e. it facilitates the evacuation of i.

The first snapshot (top-left) corresponds to the most standard situation: the agent i (black grain) is pushed by people behind them, and if they push harder, they shall accelerate i. The FiS effect is observed for people on the side: if they push harder, the contact network shall transform the additional effort into a force which pulls i away from the exit. Yet, this color distribution (standard pushers behind, counteractive pushers on the side) is not the only possibility. A second snapshot (top-right) exhibits a reversed situation. Due to the presence of a grain in front of agent i, the FiS effect is relocated in the central upstream zone, whereas red grains are swept on the sides on the jam. The two other snapshots illustrate more complex situations, where the FiS zone (blue grains) consists in several pockets distributed overall the jam.

Remark 4.8. A similar approach is proposed in Section 7.3 to the macroscopic version of the granular model. In this setting, B and $B^\star$ are replaced by their macroscopic counterpart $-\nabla\cdot$ and ∇, respectively, and $BB^\star$ is the Laplacian Δ. In spite of its infinite-dimensional character, the macroscopic setting is in some way simpler, since it involves standard differential operators which are homogeneous and isotropic. In particular, the adjoint problem on the scalar field q is a Laplace problem with Dirichlet boundary conditions -1 on the exit and 0 on the back of the congested zone (and homogeneous Neuman conditions on the wall). The gradient of the objective function is then ∇q, the direction of which is in general close to that of the desired velocity. As a consequence, no FiS effect is observed. In the present microscopic setting, the paradoxical phenomenon is made possible by the very properties of the discrete Laplace operator.

Remark 4.9. In complement to the previous remark, let us also mention a feature that is characteristic of the microscopic setting. At the macroscopic level, the definition of the objective function as the flux through the exit leads to the Dirichlet boundary condition -1 on the exit. At the microscopic level, the objective function is the velocity of an individual upfront in the desired direction. The Dirichlet condition is replaced by a right-hand side (see Eq. (4.24)) which will affect all contact points in which the grain i is involved. For each of these contacts, the corresponding entry is the opposite of the first order variation of the distance between the grains. As a consequence, the corresponding vector is likely to have both negative (which is normal in some way, and corresponds to the macroscopic setting) and *positive* entries. This situation is illustrated in Fig. 4.13. If the disk i is moved along the vector pointing to the exit, the distances with 3 and 4 will increase (i.e. negative entries) whereas, for the disks 1 and 2 on the side, this distance will *decrease*. This latter feature has no equivalent in the macroscopic setting, and this is a source of huge difference in terms of behavior between the two levels of description.

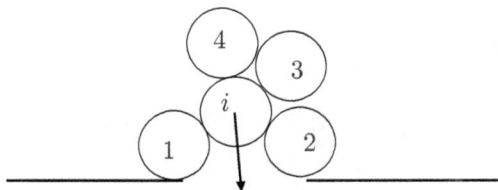

Fig. 4.13.   Illustration of Bn_i.

Chapter 5

Cellular Automata

This chapter is dedicated to a very particular class of models, namely *cellular automata*. Those models are microscopic in the sense that agents are represented individually, but they are also natively discrete in both space and time. In space: agents are identified to particles located in *cells* of a fixed cartesian grid, with an exclusion rule (one particle at most in each cell). In time: the evolution consists of a succession of *steps*, i.e. positions of particles are updated one after the other, in a stochastic manner.

Overview. The approach presented in this chapter relies on a cartesian grid, the cells of which are the only possible locations for individuals/tokens. Considering a given configuration (0 or 1 token in each cell), the model relies on a set of rules according to which individuals move, which leads to the next configuration. In the *sequential update* setting, individuals are handled one after another: each one randomly chooses a cell among the neighboring ones which are free, according to hopping probabilities which are chosen in such a way that the motion is biased in the desired direction (which depends on its individual objective and possibly other factors, like the tendency to follow, or avoid, other pedestrians). In the parallel update setting, all moves are drawn simultaneously. When all attempted moves have been determined, conflicts (i.e. 2 or more individuals aim at occupying the same cell) are listed and resolved.

5.1. Cellular Automata: General Principles

In the native version of this model, the evolution follows the following principles:

(1) *Individual tendencies.* The very notion of desired velocity is usually disregarded in this context: spontaneous trends are encoded in a so-called *Static Floor Field* (SFF), that is a discrete scalar field (one value per cell) which quantifies the attractiveness of the cell. In some situations (e.g. evacuation of a room), this field is the same for all agents, but in other contexts it may be relevant to consider different SFF for different agents. It is typically taken proportional to the distance from the closest exit. Transition (or *hopping*) probabilities are computed in such a way that the random motion is biased toward the direction of the most attractive cells.

(2) *Interactions.* Handling of interactions between pedestrians relies on two ingredients:

  (a) *Exclusion principle.* As already mentioned, there can be at most one particle on each cell. In the basic version of the algorithm, when individual moves are made sequentially, it simply means that a cell which is already occupied is not considered as a possible destination (see Fig. 5.1, center or right, the particle cannot move upward because the cell is already occupied). In more sophisticated versions of the model, desired updates are computed simultaneously, which is likely to lead to conflicts (i.e. two or more particles targeting the same cell), and those conflicts are resolved according to predetermined rules.

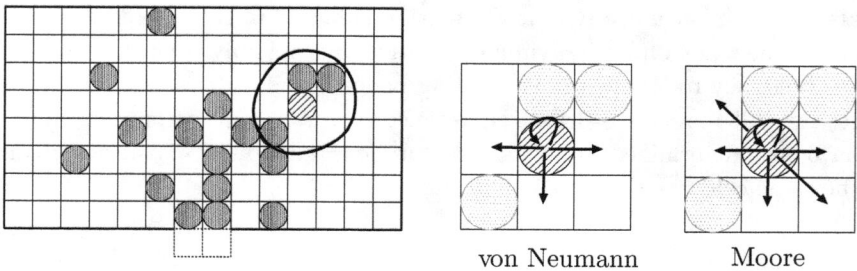

von Neumann   Moore

Fig. 5.1.   Cellular automata.

(b) *Long-range interactions.* In order to model some sort of long-range interaction, a *Dynamic Floor Field* (DFF) is sometimes added to the aforementioned SFF. It can be used to model the tendency of pedestrians to follow the overall motion of the crowd (see e.g. Schad-schneider, 2001). In that case it corresponds to a virtual trace left by the pedestrians themselves, and obeys its own dynamics, namely diffusion and decay. This field can be interpreted as some sort of chemotactic agent. Note that the very same term of DFF is used by other authors (see e.g. Hartmann *et al.*, 2014) to implement the tendency to avoid overcrowded areas. In this setting, one considers that the speed of the pedestrian depends on the local density, and this field of speeds over the room is used to determine a "modi-fied distance field", which accounts for current crowed areas, in the spirit on the approach presented in Section 8.3.

5.2. Algorithms

This general philosophy has been instantiated in many ways by various authors. It has also been followed to build industrial software, based on ingredients that are generally not made public. Recognizing the arbitrary character of our choice, we shall focus here on two versions of this model, a first one based on a *sequential* handling of events (particles are moved one after the other, in some predetermined, and possibly varying, order, see Arita *et al.*, 2015), and a second one based on a simultaneous (or *parallel*) computation of all desired updates, followed by a correction phase to handle conflicts (see e.g. Blue and Adler, 1998). We shall consider here inter-changeable individuals: updates are computing according to rules which only depend on the position, and on the type of neighboring cells (empty or occupied). Besides, we shall focus here on evacuation processes, but the approach can be straightforwardly extended to other situations (by considering alternative static floor field).

Sequential update

We first give an informal description of the time-stepping algorithm. The room is covered by a cartesian grid, the step size of which is typically 0.4 m (see Fig. 5.1, left), and the initial distribution of N "particles" respects the exclusion principle (one person/particle at most in each cell). The static floor field is defined as the distance to the exit.

One step in time consists in moving all particles sequentially according to the following rules. Suppose particles $1, 2, \ldots, i-1$, have been handled. The update of i will be determined in a stochastic way, by computing transition probabilities on neighboring cells (including the current position of i). First, the probability is 0 for cells which are occupied, considering the current configuration of the collection (i.e. accounting for the moves of the $i-1$ previously updated particles), or for cells which correspond to obstacles or walls. For the remaining welcoming cells (non-empty set: standing still is always an option), the probabilities are computed in such a way that a motion toward the direction of decreasing SFF is favored (see Eq. (5.2) below). The particle is then moved accordingly. The exit process is handled as follows: extra cells are added right downstream the exit. Those cells are accessible, and when a particle attains one of those, it immediately disappears with probability 1 during the next step.

In the *frozen shuffle update*, the indexing of individual is randomly chosen at the beginning, and remains the same over the time steps, i.e. the order in which updates are computed (which may significantly affect the overall behavior of the algorithm) is kept unchanged over all the time steps (see Fig. 5.2). Another version is proposed in Arita *et al.* (2015), called *Random Shuffle update*. The order is drawn anew at the beginning of each time step.

Formal setting

The room to evacuate is a rectangle $L_x \times L_y$, covered by a regular grid of size $\Delta x = 0.4$ m. To fix the ideas, we consider that the door is located along two adjacent edges in the middle of the bottom edge (like in Fig. 5.1). The location of an individual is encoded by an ordered pair $(i,j) \in [1, N_x] \times [1, N_y]$, with $N_x = L_x/\Delta x$, $N_y = L_y/\Delta x$. The initial configuration is given as

$$p_1^0, p_2^0, \ldots, p_N^0, \text{ with } p_\ell^0 = (i_\ell^0, j_\ell^0).$$

Let us denote by $\mathbf{x}_{ij} = ((i-1/2)\Delta x, (j-1/2)\Delta x)$ the center of cell (i,j), and by $\mathbf{x}_0$ the common target (e.g. the center of the door). The discrete Static Floor Field S is defined as

$$S_{ij} = |\mathbf{x}_{ij} - \mathbf{x}_0|.$$

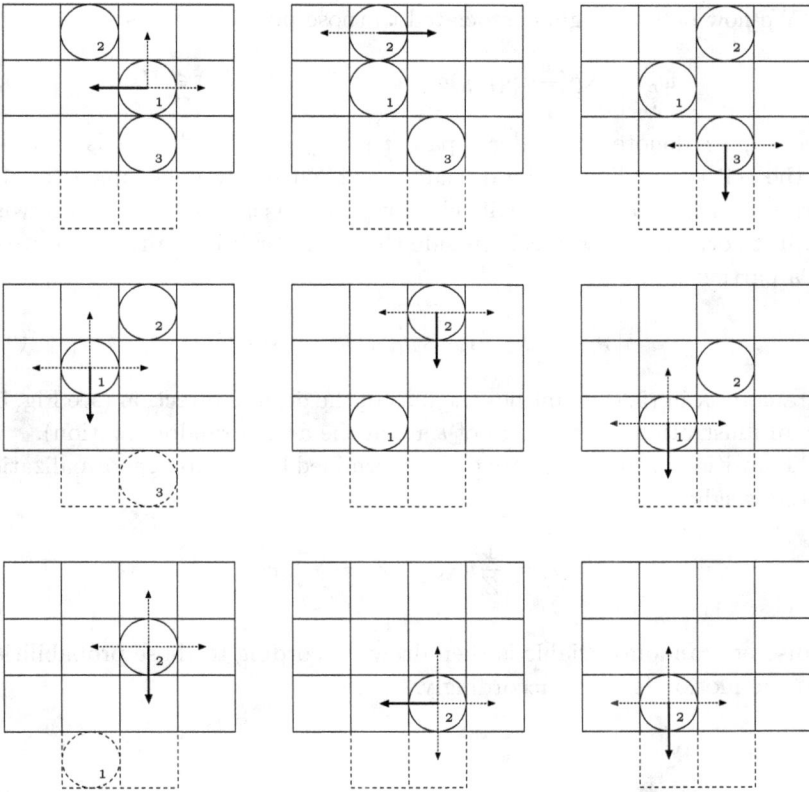

Fig. 5.2.   Frozen sequential update.

From k to $k+1$. Assume that the configuration at time step k is known[1]: $p_1^k, \ldots, p_N^k$ are given. Individual updates are performed as follows. Assume that particles $1, 2, \ldots, \ell-1$ have been updated. Particle ℓ lies at the center of a 3×3 grid. We shall write $(i,j) = (i_\ell^k, j_\ell^k)$ to alleviate notation. The core of the model consists in defining transition probabilities associated to potential moves $h \in H$, where H is the set of possible moves[2]

$$H = \{(0,0), \ (-1,0), \ (1,0), \ (0,-1), \ (0,1)\}. \tag{5.1}$$

---

[1]For particles that already left the room at time k, we shall simply disregard the corresponding indices in the procedure which follows.

[2]We favor here the von Neumann neighborhood. Choosing the Moore neighborhood consists in adding four additional moves, along the diagonal directions: $(-1,-1)$, $(-1,1)$, $(1,-1)$, $(1,1)$.

Crowds in Equations

We now define weights associated to those possible moves:

$$w_h = \exp(-\kappa S_{ij+h})\delta_{ij+h}, \qquad h = (h_x, h_y) \in H, \qquad (5.2)$$

where $ij + h$ denotes the ordered pair $(i + h_x, j + h_y)$, and δ_{ij+h} is 1 as soon as the cell $ij + h$ lies in the domain and is currently unoccupied (or if it corresponds to one of those exit cells downstream the exit), and 0 otherwise, i.e. if it corresponds to a cell outside the room, or if it is already occupied by a particle among

$$p_1^{k+1}, \ p_2^{k+1}, \ldots, p_{\ell-1}^{k+1}, p_{\ell+1}^k, \ldots, p_N^k. \qquad (5.3)$$

Parameter κ in (5.2) quantifies the bias in the desired direction (see Fig. 5.3 for an illustration of the effect of κ upon the actual random motion).

Transition probabilities are then determined by a simple renormalization of the weights:

$$p_h = \frac{1}{Z} w_h, \qquad Z = \sum_{h \in H} w_h.$$

A discrete random variable is then drawn according to those probabilities, and the motion is made accordingly.

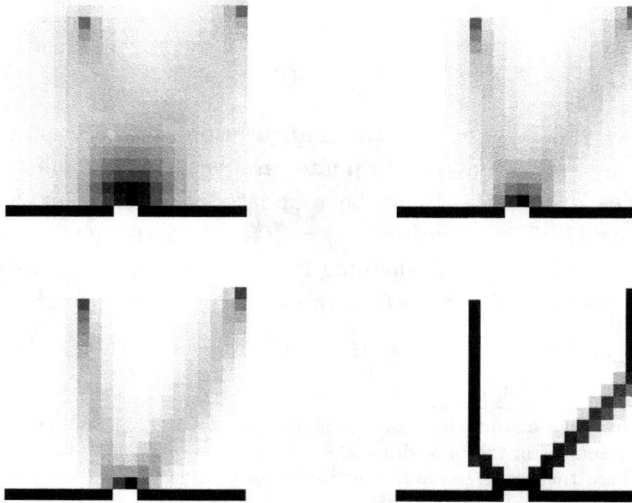

Fig. 5.3.   Influence of $\kappa = 1, 3, 5, 60 \, \mathrm{m}^{-1}$ for a 2-pedestrian evacuation.

In the *random shuffle update* situation, a new ordering is simply drawn before each time step.

Parallel update

In this setting, the evolution step is decomposed in two phases. First, *desired* moves are precomputed, i.e. for each particle, local weights are determined according to (5.2), where the local exclusion parameter δ_{ij+h} is based on positions at the previous time steps, i.e. a cell is a potential target if it lies within the room, and it is occupied by none of $p_1^k, \ldots, p_N^k$. This leads to a predicted configuration which is likely to violate the exclusion principle. Local violations are called *conflicts*, and those conflicts can be resolved as follows: if a previously empty cell is targeted by more than one agent, a winner is chosen according to the probabilities that competitors had assigned to this cell. It amounts to favor the most committed agent.

Friction

The term friction is used in Kirchner *et al.* (2002) to designate the effect induced by a modified handling of conflicts. This ingredient pertains to the parallel strategy: desired updates are drawn for all current particles according to the principles presented in the previous section, then conflicts are handled in a modified way. More precisely, one defines $\mu \in [0, 1]$ as the probability that a conflict remains unresolved (no one moves). Whenever a conflict occurs, one considers that no agent will move to the target cell with probability μ. With the complementary probability $1 - \mu$, the conflict will be resolved by allowing one of the competitor to win the conflict, like in the case without friction. Note that some authors (see e.g. Yanagisawa *et al.*, 2009) resolve the conflict by drawing a winner according to a uniform probability among competitors.

Comments on the model, tuning of parameters

Space step. The space step is usually chosen in such a way that full saturation (a person in each cell) corresponds to a realistic maximal value for the density. The value $\Delta x = 0.4\,\mathrm{m}$, which is commonly taken in the literature, corresponds to a maximal density $\rho = 6.25\,\mathrm{P\,m}^{-2}$.

Time step. The model that has been presented as a discrete succession of virtual events, without any connection yet with time. The link with a time evolution process relies on the choice of a time step Δt which separates two updates. Since the model is not the time discretization of a time-continuous process, there is no canonical way to set it up. It is nevertheless crucial to calibrate it right, in order to extract quantitative information (like evacuation times) from the computations. In the literature, it is usually taken equal to $\Delta t = 0.27$ s. Since only neighbors are considered in the jump process, with $\Delta x = 0.4$ m, it corresponds to a maximal speed of $v = \Delta x / \Delta t \approx 1.5$ m s^{-1}. Note that the stochastic character of the evolution process, together with the exclusion rule, reduce the mean speed of an individual to smaller values (see Section 5.4 for further developments on those issues).

Parameter κ, strength of attraction. The parameter κ plays a crucial role in determining the transition probabilities to neighboring cells (renormalized weights defined by (5.2)). For $\kappa = 0$, probabilities are equidistributed over the admissible cells, it corresponds to an isotropic random walk (no direction is favored). For κ growing to infinity, the model becomes deterministic in general. More precisely, if one excludes the situation where the static floor field S is affine, with a gradient aligned with one of the main diagonal directions, all weights w_h defined by (5.2) are distincts, and the associated probabilities converge to 1 for the cell $ij + h$ which maximizes the weight, and 0 for all the other cells (deterministic motion). Note that, if the maximum is attained at more than one cell, the limit transition probabilities correspond to the uniform distribution of the set of cells which realize this maximum (see Fig. 5.3, bottom-right). The model makes more sense when κ takes an intermediate value. Note that κ scales in m^{-1}, it is the inverse of a characteristic distance. Its product with S quantifies the exponential variations of the weights. If one considers that S is the distance to a target, the variations of S_{ij+h} over the potential destinations scale like Δx. As a consequence, for κ significantly smaller than $1/\Delta x$ the random walk will essentially be isotropic, whereas for κ significantly larger than $1/\Delta x$, the walk will be deterministically biased in the most favorable direction. A reasonable value is $\kappa \approx 1/\Delta x$. The value of κ not only quantifies the free speed of individuals, but also quantifies the diffusion in the transverse direction (orthogonal to the desired direction). To illustrate those considerations, we plot in Fig. 5.3 the dispersion of trajectories for two pedestrians evacuating a room. More precisely, we computed several evacuation scenarios for different values of κ (same initial condition, different stochastic trials). For each value of κ, we plot the corresponding presence density function on the underlying grid: the grey level corresponds to the number of times each cell has been visited during all the computations.

Friction. The friction parameter $\mu \in [0,1]$ is the probability that, in the event of a conflict (two or more particles compete for the same cell), none of the protagonists move. The no-friction case thus corresponds to $\mu = 0$: each conflicts ends up with an actual motion. On the opposite, $\mu = 1$ would tend to freeze evacuation of pedestrians toward a common destination (e.g. the exit of a room). Again, reasonable choice lie in-between. Compared numerical experiments are proposed in Yanagisawa *et al.* (2009), for $\mu = 0.25$ and 0.18. Higher values $\mu = 0.6$ and $\mu = 0.9$, are tested in Schadschneider *et al.* (2009), making it possible to recover the so-called Faster-is-Slower effect (see Section 9.7). We refer also the reader to Section 10.1 for an elementary computation explaining how this parameter tends to affect the overall efficiency of an evacuation process.

To illustrate the effect of this friction parameter, we represent in Fig. 5.4 snapshots of an evacuation of 200 agents, at discrete times 0, 100, 200, and 250. The gray level reflects the initial distance to the exit. The evacuation is less efficient for larger friction parameters.

Fig. 5.4. Friction parameter $\mu = 0$, 0.25, 0.5, 1, respectively.

5.3. Variations, Extensions

Various ingredients have been proposed in the literature to enrich the basic models presented in the previous section.

Dynamic floor field

The term Dynamic Floor Field (DFF) designate a discrete scalar field (one value for each cell) which is used to account for non-local effects on the tendencies of agents. The effect that is implemented may vary from a model to another.

In the approach proposed in Schadschneider (2001), it can be interpreted as a virtual trace left by moving agents, it is meant to encode the attraction tendency between pedestrian. It plays the role of a chemoattractant, like in the Keller–Segel system (Keller and Segel, 1970). In the latter setting, the moving entities are cells (e.g. *E. coli*). Those cells generate a substance which diffuses in the extracellular medium, decaying progressively, and tend to *attract* the cells themselves. This attraction is usually included in the model by adding an advection term in the equation modeling the cell motion, with a velocity taken proportional to the gradient of the chemoattractant. In the context of crowd motion, this signal is no longer chemical, but it can be used to model the tendency of pedestrians to head toward zones which have been recently visited. Like in the Keller–Segel context, the evolution is based on three mechanisms:

(1) **Creation**. The presence of an agent in a cell tends to increase the local value of the field.
(2) **Diffusion**. A certain amount of the available field is distributed among the neighboring cells.
(3) **Decay**. The value decays at a predetermined rate.

At a continuous level in time and space, such a process follows a reaction diffusion equation of the type

$$\frac{\partial D}{\partial t} - \Delta D = -\mu D + \beta P,$$

where P is the local density of agents (each of which emits the virtual substance at a constant rate). At the discrete level, the evolution reads

$$D_{ij}^{k+1} = (1 - \mu\Delta t)D_{ij}^k + \beta P_{ij}^k$$
$$+ \beta(D_{i-1,j}^k + D_{i+1,j}^k + D_{i,j-1}^k + D_{i,j+1}^k - 4D_{ij}^k),$$

where $P_{ij}^k = 1$ when the cell is occupied at time step k, and 0 otherwise.

Enriched neighborhoods

In the standard version of the CA model, the only authorized motions are up, down, left, and right. Richer neighborhoods are considered in Kretz and Schreckenberg (2006), and their effects upon the resulting average speed of discrete agents are quantified.

5.4. Cellular Automata, Mathematical Issues

Cellular automata were introduced to investigate the behavior of multiparticle systems. The core of the approach lies in the simplicity of the microscopic evolution rules (motion/creation/suppression of particles), which makes it possible to straightforwardly perform computations, and to elaborate theoretical results on the overall behavior of the discrete systems. In the present case of crowd motion, the exclusion principle fits in the standard framework of CA, but the rule is no longer the same for all cells. This is due to the Floor Field (static or dynamic) which, except in very simple situations, does not induce uniform transition probabilities. This ingredient made the approach applicable to crowds, but it also rules out most of the standard theoretical tools, making their theoretical study a very delicate matter.

We shall restrict ourselves to properties which can be established in very particular situations, to enlighten the methodological core of the method, and also to help choosing the various parameters or algorithmic choices, depending on the phenomena we aim at reproducing.

Mean free speed

Consider a single particle in a one-dimensional grid with step size Δx, submitted to the Static Floor Field $S(x) = x$ (the particle tends to minimize its abscissa, i.e. to move leftward). We assume that hopping probabilities are defined by the one-dimensional version of (5.2). The associated Cellular automaton is an homogeneous random walk on $\mathbb{Z}$, with transition probabilities

$$\pi_{-1} = \frac{1}{Z}\exp(\kappa\Delta x), \quad \pi_0 = \frac{1}{Z}, \quad \pi_{+1} = \frac{1}{Z}\exp(-\kappa\Delta x),$$

$$Z = \sum_{m=-1}^{+1}\exp(-m\kappa\Delta x).$$

The mean velocity $\overline{V}$, that is the expected displacement divided by the time step, can be straightforwardly computed:

$$\overline{V} = -2\frac{\Delta x}{\Delta t}\frac{\sinh(\kappa\Delta x)}{1 + 2\cosh(\kappa\Delta x)}.$$

As expected, $|\overline{V}|$ is in $[0, \Delta x/\Delta t]$ (the maximal displacement during Δt is Δx).

This expression can be used to calibrate the value of κ in order to recover a prescribed velocity U. If one considers the standard setting $\Delta x = 0.4\,\text{m}$, $\Delta t = 0.27\,\text{s}$, the values range between 0 and $\Delta x/\Delta t \approx 1.48\,\text{m\,s}^{-1}$. For $\kappa = 3\,\text{m}^{-1}$, the corresponding speed is around $U = 1\,\text{m\,s}^{-1}$. For smaller values of κ, one can use the first-order expansion of the previous expression to set the value κ which corresponds to a prescribed speed U:

$$\kappa \approx \frac{3}{2}\frac{\Delta t}{\Delta x^2}U.$$

Sequential update: frozen versus random update

We consider a one-dimensional version of the cellular automaton in the sequential version. To focus on the influence of the order in which updates are made, we simplify the situation by assuming a purely deterministic behavior: when its turn comes, each particle makes a one-cell move in the right direction, with probability 1. Note that it corresponds to the asymptotic $\kappa \to +\infty$. We consider a fully congested initial configurations: N adjacent cells are occupied by the same number of particles. We first consider the frozen setting: the initial ordering is drawn randomly. We represent this ordering from the right to the left: p_1 is the head particle, then p_2 the one that lies on its left, up to p_N. In the first update phase, a cluster of size m_1 will separate and move one step rightward, where $m_1 \geq 1$ is the largest integer such that

$$p_1 < p_2 < \cdots < p_{m_1}.$$

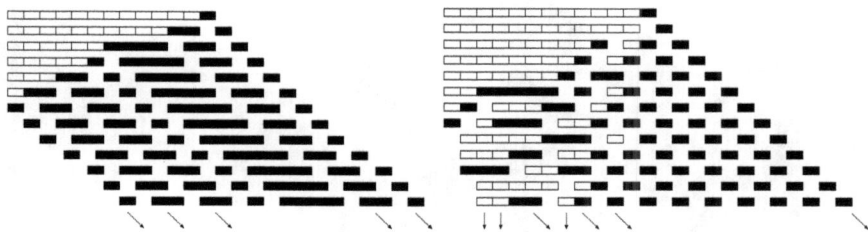

Fig. 5.5. Frozen (left) and random (right) shuffle updates.

This size m_1 is equal to 1 in the example presented in Fig. 5.5 (left). Since the ordering remains unchanged, this cluster will continue its rightward motion at pace 1 cell per update step. The remaining $N - m_1$-cluster will be separated into a moving one on its head, of size m_2, where m_2 is defined in a similar way ($m_2 = 2$ in the example), etc. The description of the distribution of cluster sizes is a delicate matter, which has generated an important bibliography (fully disconnected from the study of the present algorithm). It can be shown in particular that the expected size of a cluster is asymptotically 2, but the expected size of the kth cluster oscillates around 2 when k increases, with a period close to 5.3, as proven in Hooker (1969) using complex-variable theory.

As for the *random* shuffle update, the first step is identical, but there is a main difference: since the ordering changes at each iteration, the local ordering within each head cluster is likely to change, so that each cluster will tend to further separate into two sub-clusters, until all clusters are decomposed onto atoms (single particles). A typical evolution is presented in Fig. 5.5 (right).

Note that, in both situations, the head particle moves at unit constant speed. In the randomized case, the evolution tends to form smaller and smaller clusters, down to isolated particles, whereas in the frozen case, clusters are sustainable, and the final configuration is characterized by a coarser granularity.

Chapter 6

Compartment Models

The models presented in this chapter are fundamentally different from all other models in this book. They do not aim at computing positions and velocities, but rather at describing the evolution of total amounts of people in specified *compartments*. The compartments shall be identified to *nodes* of a network, those nodes are connected by *edges*, which correspond to paths between nodes. Although they are both macroscopic and Eulerian, they rely on a finite number of degrees of freedom, and therefore can be understood and studied quite straightforwardly. They prove to be very efficient in some situations, in particular when the path of individuals is determined in advance (like in emergency evacuations, under the assumption that each individual follows the shortest path to the set of exits). We shall favor in our presentation the standpoint of emergency evacuation. Let us add that, in spite of their conceptual simplicity, properly writing those models is delicate from the mathematical standpoint. For this reason, we shall favor an informal presentation of the models together with a description of numerical algorithm, and put off until the end of this chapter the underlying mathematical issues. A full understanding of the mathematical framework is not mandatory to grasp the model itself and its effective numerical implementation.

6.1. Compartment Models: Toy Versions and General Setting

Let us start with the situation represented in Fig. 6.1 (left). A certain quantity of people is accumulated upstream an exit door. Since we aim

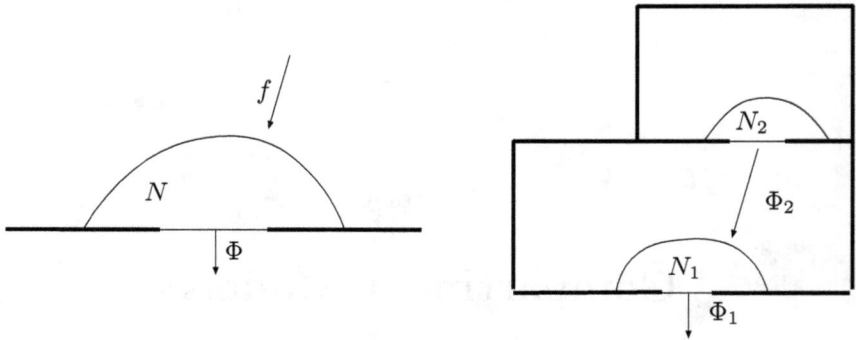

Fig. 6.1.   One- and two-room toy problems.

at setting a continuous model, applicable to large number of entities, we
represent this quantity by a *real number* $N \in \mathbb{R}_+$. The capacity of the exit,
that is the maximal number of individuals which can go through it per
unit time (see Section 9.5), is denoted by C. We shall assume here that
C is constant, and we refer to Section 6.3 for extensions. We denote by
$f = f(t) \geq 0$ the incoming flux, that is the upstream flux of pedestrians,
and by $\Phi = \Phi(t) \geq 0$ the instantaneous flux through the door. The balance
at the door writes $dN/dt = f - \Phi$. The core of the model relies in the
expression of Φ as a function of the dynamic variables N, f, and the static
parameter C. By definition of the capacity it holds that $\Phi \in [0, C]$, and
$\Phi = C$ whenever $N > 0$. When $N = 0$, Φ lies between 0 and C. Its value is
f when $f < C$, but it may happen that Φ saturates to C if $f > C$.

To sum up, the evolution problem can be written

$$\begin{cases} \dfrac{dN}{dt} = f - \Phi, \\[2ex] \Phi = \begin{cases} C & \text{if } N > 0, \\[1ex] \arg\min\limits_{\tilde{\Phi} \in [0,C]} |\tilde{\Phi} - f| & \text{if } N = 0. \end{cases} \end{cases} \tag{6.1}$$

Extension to two rooms is straightforward. Consider the situation represented in Fig. 6.1 (right): the accumulation upstream the exit of room 1 is
supplied by an influx of people coming from room 2. We denote by T_1 the
time need to walk from this exit of 2 to the exit of 1 (we neglect the size of
the jam itself). We assume that all people initially gathered in room 2 are

gathered near the exit, we obtain the following system:

$$
\begin{cases}
\dfrac{dN_2}{dt} = -\Phi_2, \\[2mm]
\dfrac{dN_1}{dt} = -\Phi_1 + \Phi_2(t - T_1), \\[2mm]
\Phi_2 = \begin{cases} C_2 & \text{if } N_2 > 0, \\ 0 & \text{if } N_2 = 0, \end{cases} \\[4mm]
\Phi_1 = \begin{cases} C_1 & \text{if } N_1 > 0, \\ \arg\min\limits_{\tilde{\Phi}\in[0,C_1]} |\tilde{\Phi} - \Phi_2(t - T_1)| & \text{if } N_1 = 0. \end{cases}
\end{cases}
\tag{6.2}
$$

where Φ_i stands for $\Phi_i(t)$ (we only express the time dependence when the value is taken at a time in the past, like for $\Phi_2(t - T_1)$).

General case

We shall present the many-compartment situation in a particular situation. We consider the one-floor building represented in Fig. 6.2, with six rooms, each of which contains a single exit. The underlying structure is represented by a directed graph with six vertices (each of which corresponds to the exit of the corresponding room). We choose the orientation in such a way that arrows stemming from a vertex i point to the vertices that influence i (upstream orientation). For the example of Fig. 6.2, the arrows of the

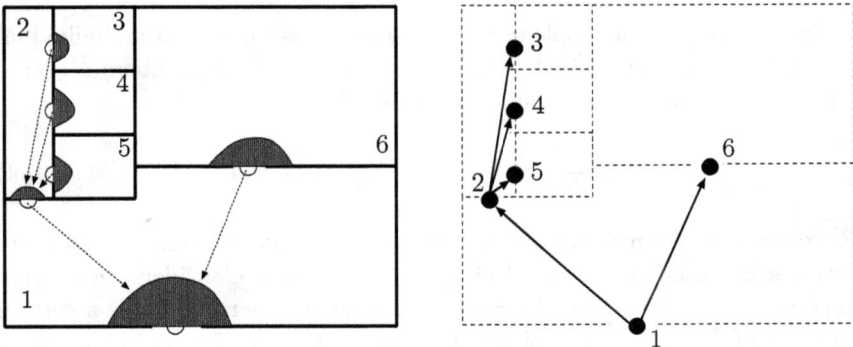

Fig. 6.2. Complex building and its associated oriented network.

graph are

$$1 \longrightarrow 2, \ 1 \longrightarrow 6, \ 2 \longrightarrow 5, \ 2 \longrightarrow 4, \ 2 \longrightarrow 3.$$

We denote by m_i the number of inlet accesses to room i. For $n = 1, \ldots, m_i$, we denote by α_i^n the index of the room upstream the access n to room i. The time spent by an individual to walk (in room i) from entrance n to the exit is T_i^n. Finally, Φ_i is the instantaneous flow rate of people through the exit of i.

The problem writes

$$\begin{cases} \dfrac{dN_i}{dt} = -\Phi_i + \displaystyle\sum_{n=1}^{m_i} \Phi_{\alpha_i^n}(t - T_i^n), \quad i = 1, \ldots, 6, \\[4mm] \Phi_i = \begin{cases} C_i & \text{if } N_i > 0, \\[3mm] \arg\min\limits_{\tilde{\Phi} \in [0, C_i]} \left| \tilde{\Phi} - \displaystyle\sum_{n=1}^{m_i} \Phi_{\alpha_i^n}(t - T_i^n) \right| & \text{for } i = 1, \ldots, 6. \end{cases} \end{cases} \quad (6.3)$$

Full account of people in the building

The approach makes it possible to follow also the headcount of people who are not gathered upstream any door. More precisely, we shall denote by $\overline{N}_{i,n}$ the number of people on their way to the exit of room i, coming from the exit of room α_i^n, so that

$$N_i + \sum_{n=1}^{m_i} \overline{N}_{i,n}$$

is the total number of people in room i. Since we assumed that all individual were initially gathered right upstream exit doors, those numbers $\overline{N}_{i,n}$ are set to 0 at $t = 0$. Their evolution is governed by

$$\frac{d\overline{N}_{i,n}}{dt} = \Phi_{\alpha_i^n}(t) - \Phi_{\alpha_i^n}(t - T_i^n). \quad (6.4)$$

Remark 6.1 (Communicating vessels). This model can be seen as an extreme instance of a class of simple fluid models. Consider a collection of containers as illustrated by Fig. 6.3. Each container i contains a certain quantity of fluid which is proportional to the height h_i. The fluid flows from 1 to 0, from 2 to 0, and from 0 to the outside world. Now consider that the hole at the bottom of each container is actually a narrow pipe, with

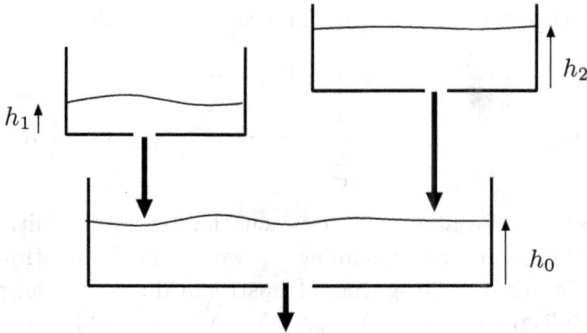

Fig. 6.3.   Fluid compartment model.

predominant viscous effect. The flow is then proportional to the pressure upstream, which is proportional to the height, so that the flux Φ_i can be expressed as Ch_i. The global problem takes the form of (6.3), where N_i is replaced by h_i, and the flux are explicitly expressed in terms of the h_i's. Now consider that inertial effects are predominant compared to viscous one, to the point that the fluid behaves like a perfect fluid obeying Bernoulli's law. The flux can then be written as proportional to the square root of the hydrostatic pressure right upstream the holes, which scales like the height: we obtain again a problem similar to (6.3), with $\Phi_i = C\sqrt{h_i}$. The crowd evacuation case can therefore be seen as a singular limit of generalized fluid models, by considering expressions of the type $\Phi_i = Ch_i^\alpha$, with $\alpha \to 0$.

6.2.   Numerical Solution

In spite of non-trivial mathematical issues (see Section 6.5), designing numerical algorithms to solve this sort of problems is straightforward. Let $\tau > 0$ denote a time step. To simplify the presentation, we shall assume that the transfer times are whole multiples of this time step, and denote by $\tilde{T}_i = T_i/\tau \in \mathbb{N}$ the corresponding dimensionless times. We shall furthermore assume that people are initially gathered in the neighborhood of exits. The approximation of N_i at time $t^k = k\tau$ is denoted by N_i^k.

The scheme is actually simpler than the continuous model, since it simply expresses that, at some time t^k, the number of individual walking through exit i between t^k and t^{k+1} is $C_i\tau$ whenever there is enough people available to achieve this full capacity, and N_i otherwise. In other word, the average flux in this time interval is either C_i or N_i/τ.

$$\begin{cases} \Phi_i^{k+1} = \min(C_i, N_i^k/\tau), \quad i = 1, \ldots, R, \\ N_i^{k+1} = N_i^k - \tau \, \Phi_i^{k+1} + \sum_{n=1}^{m_i} \tau \, \Phi_{\alpha_i^n}^{k+1-\tilde{T}_i^n}, \quad i = 1, \ldots, R, \end{cases} \tag{6.5}$$

where Φ_i^k is set to 0 for all $k < 0$ (no evacuation before the initial time), and R is the number of rooms. Assuming, as we did in the continuous setting, that all people are initially gathered upstream doors, we supplement this system with initial conditions $N_1^0, \ldots, N_R^0$. We may add an extra equation to keep an account of people who have evacuated the building at time k:

$$N_0^{k+1} = N_0^k + \tau \, \Phi_1^{k+1}.$$

It is also straightforward, for each room i, to keep an account of the number $\overline{N}_{i,n}$ of individuals who are on their way to the exit of room i, coming from the exit of room α_i^n, by discretizing Eq. (6.4):

$$\overline{N}_{i,n}^{k+1} = \overline{N}_{i,n}^k + \tau \Phi_{\alpha_i^n}^{k+1} - \tau \Phi_{\alpha_i^n}^{k+1-\tilde{T}_i^n}.$$

Global people balance is straightforwardly obtained by summing up all discrete equations:

$$N_0^k + \sum_{i=1}^R N_i^k + \sum_{i=1}^R \sum_{n=1}^{m_i} \overline{N}_{i,n}^k = \text{Constant}.$$

In particular, if

$$N_{\text{tot}}^0 = \sum_{i=1}^R N_i^0$$

is the total number of people initially present, and if all capacities are positive, for k sufficiently large (i.e. such that $k\tau$ is larger that the evacuation time), it can be checked that N_i^k is 0 for $i = 1, \ldots, R$, and $N_0^k = N_{\text{tot}}^0$ (everybody left the building).

6.3.   Extensions

Capacity drop

As mentioned in the introduction, accounting for the capacity drop effect (see Section 9.6) is possible, as soon as one considers as valid a given expression of the capacity C with respect to the number N of people accumulated upstream the door. The mapping $N \mapsto C(N)$ would be typically non-increasing. The non-increasing character affects the nature of the model from the mathematical standpoint (see Section 6.5), but it does not change either the informal presentation we made, nor the numerical strategy. Nevertheless, would this dependence of C with respect of N be discontinuous at some point N_c, the evolution may present a highly instable behavior in the neighborhood of this regime (depending on whether the current N is slightly larger or smaller than N_c). The expected behavior is emphasized by a loss of uniqueness of the underlying evolution model.

Lagrangian tracking of individuals

The compartment approach relies of quantities and fluxes of people defined at some predefined zones: it is *Eulerian* in nature. Yet, it makes it possible to build a posteriori the paths of single individuals, as an extra outcome of the computed evolution, given some assumptions are made. The crucial assumption is of the FIFO (First In First Out) type: an individual joining a jam at some time t will exit the room *after* all individuals already in the jam at t, and *before* all who were not yet in at time t. Consider for instance an individual initially located at the exit of room 5 (see Fig. 6.2). He/she reaches the exit 2 at the time that is needed to travel between 5 and 2, that is T_2^1, with $\alpha_2^1 = 5$. At this very instant, the jam contains $N_2(T_2^1)$, so that it will take $N_2(T_2^1)/C_2$ to exit room 2 and enter room 1, at time $T_2^1 + N_2(T_2^1)/C_2$, and so on.

Expanding bifurcation

The approach presented above applies to evacuation situations, or more generally to situations where possible paths of individuals entering a room is unique. All bifurcation points we considered had consequently a *gathering* character: fluxes coming from various sources are added at the next exit. This assumption makes the approach quite straightforward from the modeling standpoint, and ruling it out is obviously likely to raise some additional issues (see Remark 6.4 on this matter). Yet, more general situations can be considered, at the price of extra assumptions on people behavior. For example, one may consider the case where the last room before the outside world (Room 1 in Fig. 6.2) has two exits. An individual entering this room has to choose between both exits, which calls for extra modeling assumptions. In the case of evacuation, assuming full rationality of agents, it would be

natural to consider that someone entering the room instantaneously estimates the costs associated to all choices: for any possible exit, estimate the
time needed to reach it, plus the time to evacuate the jam that is currently
present (which calls for a knowledge of the corresponding capacity), and
choose the cheapest solution in terms of evacuation time. One may also at
this point consider the role of signage, possibly dynamic signage, to favor
one particular choice. In more general situations, like people entering a
large building, an exhibition center, a railway station, etc, a proper modeling calls for a prior knowledge of people intentions, and signage can again
play an important role in influencing the choices of agents.

6.4.   Numerical Illustration

To illustrate the capability of this approach to rapidly compute evacuations in complex buildings, we present some tests based on the geometry
presented in Fig. 6.2. The left column of Fig. 6.4 presents the computed
evolution at times $0, 5, 10, 15$, and 20 s. The area of the half disk attached to
each of the exits is proportional to the number of people gathered upstream
this exit, whereas the elongated rectangles which link transit points have
an area which is proportional to the number of people on their way from
an exit to the next one. The column on the right is meant for comparison
purpose: it represents snapshots of the same evacuation scenario computed
with the granular model presented in Chapter 4.

6.5.   Mathematical Framework: A Cascade
of Gradient Flows

The mathematical framework relies on a reformulation of the problem as a
differential inclusion. Let us first consider the simplest toy problem: evacuation of a room through a single exit door. Like in Section 6.1, we denote
by N the number of people accumulated upstream the exit, and by f the
rate of people arriving at the exit. If the door is closed, we simply have
$dN/dt = f$. Now consider that door is open, and characterized by its (constant) capacity $C > 0$, that is the maximal number of individual who may
go through it, per unit time. Whenever N is positive, the flux Φ through
the door is C. When N is 0, all we know about Φ is that it is between 0
and C. This can be expressed by the differential inclusion

$$\Phi \in -\partial\varphi(N) = -\{v, \quad \varphi(N) + vh \leq \varphi(N + h), \ \forall h\},$$

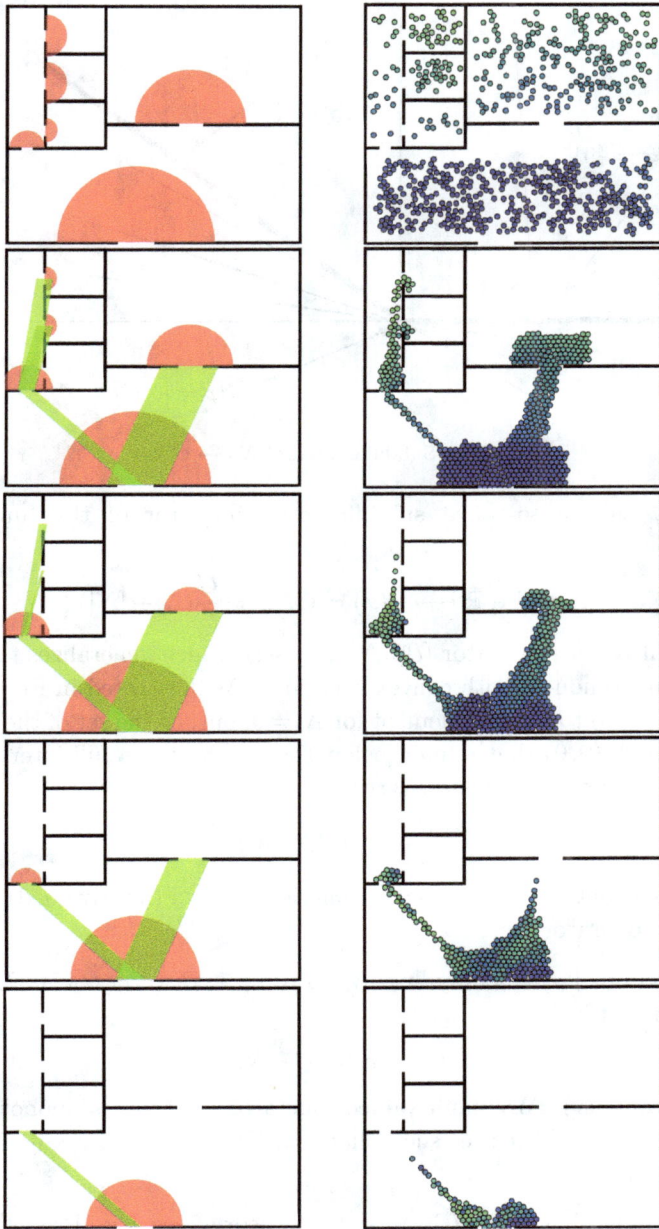

Fig. 6.4.   Compartment model vs. granular model.

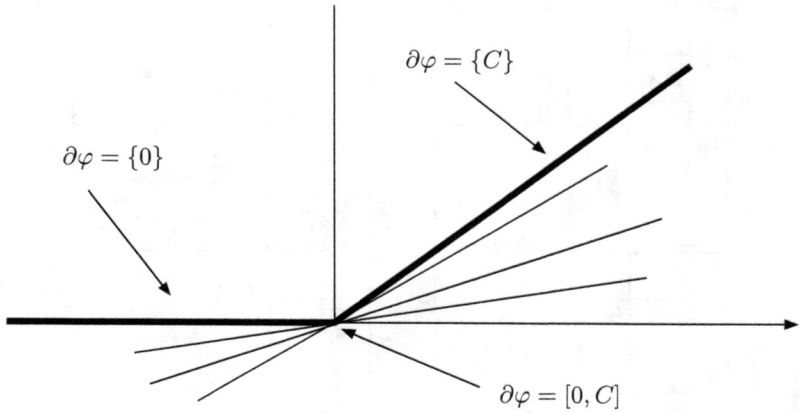

Fig. 6.5.   Subdifferential of $N \longmapsto CN_+$.

where $\partial\varphi$ is the so-called subdifferential operator of the function φ, defined by

$$\varphi \ : \ N \in \mathbb{R} \longmapsto \varphi(x) = CN_+ = \frac{C}{2}(N + |N|). \qquad (6.6)$$

It is a multivalued operator ($\partial\varphi(N)$ is a *set*), which generalizes the notion of gradient to non-smooth convex functions. As illustrated in Fig. 6.5, this set is restricted to a single element for $N \neq 0$, but, as the set of the slopes of lines through $(0,0)$ that remain below the graph, it is a full interval $[0, C]$.

Mass balance may then be written

$$\frac{dN}{dt} + \partial\varphi(N) \ni f.$$

Since φ is a continuous and convex function, $\partial\varphi$ is a (multivalued) maximal monotone operator, i.e.

(i) (Monotonicity) For any N_1, N_2, any $v_1 \in \partial\varphi(N_1)$, $v_2 \in \partial\varphi(N_2)$, it holds that

$$(v_2 - v_1)(N_2 - N_1) \geq 0.$$

(ii) (Maximality) Any multivalued operator A that is monotone and "larger" than $\partial\varphi$ (i.e. such that $\partial\varphi(N) \subset A(N)$ for any N) is $\partial\varphi$ itself.

If we assume that f is L^∞, standard theory (see e.g. Proposition 3.4 in Brezis (1973)) ensures existence and uniqueness of a solution, and this solution is Lipschitz continuous in time. This setting does not make it necessary to explicitly express the flux, but we shall need it for the multi-compartment version. The same theorem ensures existence of a L^∞ flux Φ such that

$$\frac{dN}{dt} = f - \Phi,$$

and the flux Φ is defined by

$$\Phi = \arg\min_{\tilde\Phi\in\partial\varphi(N)} |\tilde\Phi - f|. \tag{6.7}$$

Note that the latter expression elegantly combines both terms of the alternative in (6.1). Indeed, whenever N is positive, the subdifferential is reduced to $\{C\}$, so that the arg min above yields $\Phi = C$.

In the case $f = 0$, the evolution is the pure gradient flow associated to the positive part function (up to a multiplicative constant C), and the solution is explicitly given by

$$t \longmapsto \left(N^0 - Ct\right)_+,$$

where N^0 is the initial number of people accumulated upstream the exit.

Remark 6.2. The strictly monotone character of the evolution makes it irreversible, and the well-posedness result does not extend to the backward problem. This irreversibility is clear from a modeling standpoint. For example, starting with $N^0 = 0$, the only solution is the stationary solution. But considering the time-reversed situation, with a room that is initially empty, is it possible to infer the evacuation which may have taken place in the recent past? The answer is obviously negative, since the room may have been empty forever in the past, or may have been emptied some arbitrary time ago.

General setting

In the general, many-compartment situation (like in Fig. 6.2), the problem can be reformulated in the same spirit:

$$\begin{cases} \dfrac{dN_i}{dt} + \partial\varphi_i(N_i) \ni \displaystyle\sum_{n=1}^{m_i} \Phi_{\alpha_i^n}(t - T_i^n), \quad i = 1,\ldots,R, \\[4mm] \Phi_i(t) = \arg\min\limits_{\tilde\Phi\in\partial\varphi_i(N_i)} \left|\tilde\Phi - \displaystyle\sum_{n=1}^{m_i} \Phi_{\alpha_i^n}(t - T_i^n)\right| \quad \text{for } i = 1,\ldots,R \end{cases} \tag{6.8}$$

with

$$\varphi_i \; : \; x \in \mathbb{R} \longmapsto \varphi(x) = C_i \, x_+ = \frac{C_i}{2} \left(x + |x| \right).$$

Remark 6.3. The model takes the form of a *cascade* of forced gradient flows, i.e. a collection of gradient flows which are connected in a hierarchical way. If one considers that nodes are indexed in such a way that the oriented graph structure is preserved, i.e. $i \to j$ implies $i \leq j$, then the ith gradient flow has a source terms which comes from various j's, with $i < j$, taken at some times in the past. We may actually write the overall evolution model as a global gradient flow with a forcing term which depends on the strict past: introducing

$$N = (N_1, \dots, N_R), \qquad \varphi(N) = (\varphi_1(N_1), \dots, \varphi_R(N_R)),$$

we obtain

$$\frac{dN}{dt} + \partial\varphi(N) \ni \Psi,$$

where the right-hand side Ψ is built from the evolution problem itself at some previous times. More precisely, Ψ depends on the solution itself on the time interval $[0, t - T_{\min}]$, where $T_{\min}$ is the minimal transfer time.

Remark 6.4. In the spirit of Remark 6.2, the backward problem raises another kind of uniqueness issue. Consider an exit supplied by two fluxes f_1 and f_2 (computed as delayed fluxes through two entrances of the considered room). In the forward evacuation problem, mass conservation expressed at this node simply sums up the two incoming fluxes. Now for the backward problem, the very same mass balance only requires $f_1 + f_2$ to have a prescribed value (that is the flux of people "entering through the exit"), but does not say anything about the relative distributions. If one may dare express the situation in baby language, the differences in terms of well-posedness, due to a lack of monotonicity of the backward operator, is essentially due to the fact that the problem $a + b = c$ is a well-posed problem if a and b are given and c is the unknown, but ill-posed if c only is given, and a and b are the unknowns. This feature can be put in parallel with the heat equation, that is the archetypal example of evolutions problems driven by a maximal monotone operator. It is common knowledge that the (mathematical) entropy of the solutions to this equation decreases. If one considers the solution as the time-evolving probability density of a brownian particle, the decreasing character of the entropy expresses that the *quantity of information* on the actual location of the particle decreases. In the present case, we recover this feature in a caricatural way: while adding a and b to get c, a part of the information on a and b is lost.

Capacity drop

As detailed in Section 6.3, the capacity drop effect (see Section 9.6) can be modeled by having the capacity C depend on the current number of accumulated people N. In the gradient flow framework, this amounts to replace CN_+ in (6.6) by an antiderivative $\varphi(N)$ of the function $N \mapsto C(N)$. By the very nature of the capacity drop phenomenon, $C(N)$ is typically non-increasing, which make the antiderivative φ *concave*. This induces new issues in terms of well-posedness, because the subdifferential of a concave function is not defined in general (the set may be empty). This difficulty can be overcome by replacing the notion of subdifferential by a more general one, namely the Fréchet subdifferential, defined as

$$\partial \varphi(N) = \{v, \quad \varphi(N) + vh \leq \varphi(N+h) + o(h)\}$$

which straightforwardly applies to *smooth* non-convex functions. At a regular point of such a function, the subdifferential identifies with the standard derivative (or gradient in higher dimensions). The theory extends to this situation (see Brezis, 1973): a sufficient condition for well-posedness is that the operator is maximal monotone up to addition of a Lipschitz term. If the function is smooth (with a second derivative that is bounded from below), it is possible to make it convex by adding a sufficiently strong quadratic term.

The situation becomes more critical if $C \mapsto C(N)$ is discontinuous, which makes sense from the modeling standpoint since the Capacity Drop phenomenon is commonly referred to as a *sudden* drop of the capacity whenever the size of the jam upstream the exit reaches some critical N_c. At N_c the potential φ is no longer smooth, more precisely it presents a concave angular point. The theory then no longer applies, because this singularity cannot be straighten out by addition of any smooth convex correction. Indeed, uniqueness is lost. This defect in the theory reflects an instability of the evolution, which is clear from a modeling standpoint: if the capacity drop takes place at some specific point N_c, the evolution will strongly depend on the exact position of N with respect to N_c. Let us add that it strongly questions the very modeling approach, it actually suggests that the description of the phenomenon as a drop which happens above a very specific and precise value does not make much sense.

Chapter 7

Toward Macroscopic Models

We present in this chapter some macroscopic models which have been proposed to model crowd motions. There is a huge current activity on developing new models on this type and, since this book is focused on microscopic description, we shall merely focus on models which are macroscopic counterparts of the microscopic models we presented in previous chapters. This chapter is not intended for specialists in Partial Differential Equations, it is rather meant as an introduction to macroscopic modeling in the context of crowd motion, in direct connection with microscopic approaches.

Macroscopic models are based on a diffuse description of the crowd, represented as a local density $\rho(x,t)$, transported by an instantaneous velocity field $u = u(x,t)$ in a conservative way:

$$\frac{\partial \rho}{\partial t} + \nabla \cdot (\rho u) = 0, \tag{7.1}$$

where $u(x)$ is the velocity at time t of pedestrians located in the neighborhood of x. The very fact that we consider this local velocity as well-defined is a strong assumption, it requires in particular that all individuals located around x go at the same velocity. The latter assumption may be ruled out in some real-life situations, in particular when the crowd is a mixture of pedestrians with various objectives. This limit might be overcome by considering multiphase flows, based on different densities $\rho_1, \ldots, \rho_P$, where P is the number of distinct populations (i.e. with distinct objectives), but this approach raises delicate issues which we shall not address here. Coming back to our "behavioral homogeneous" assumption, modeling the motion amounts to set up rules to determine the instantaneous velocity field $u(x,t)$.

The simplest macroscopic crowd motion model one may think of, by assuming that a field of desired velocity U is given (e.g. computed according to the principles described in Chapter 8), is obtained by disregarding any interaction between pedestrians: the density ρ is then simply transported by U:

$$\frac{\partial \rho}{\partial t} + \nabla \cdot (\rho U) = 0.$$

7.1. One-Dimensional Macroscopic Traffic Model

The Follow-the-Leader model presented in Section 2.1 has a direct macroscopic counterpart, which is called LWR model (from Lighthill and Whitham (1955) and Richards (1956)). It was initially meant to model car traffic, but its relevance for crowds of people walking one behind another is obvious. It is based on the one-dimensional version of Eq. (7.1), with an advection velocity written as a function $u(\rho)$ of the local density:

$$\partial_t \rho + \partial_x \left(\rho u(\rho) \right) = 0, \tag{7.2}$$

or, written as a general conservation law:

$$\partial_t \rho + \partial_x \left(f(\rho) \right) = 0,$$

where $f(\rho) = \rho u(\rho)$ is the flux.

Propagation of perturbations. If one considers a uniform stationary solution $\rho \equiv \rho_{\text{eq}}$ of this equation, and a perturbed solution $\rho_{\text{eq}} + \tilde{\rho}$, we formally obtain an equation on the perturbation

$$\partial_t \tilde{\rho} + f'(\rho_{\text{eq}}) \, \partial_x \tilde{\rho} = 0 \tag{7.3}$$

which expresses that small perturbations around a constant state solution are transported at speed $f'(\rho_{\text{eq}})$.

Assume now that $\rho(x,t)$ is a regular solution of Eq. (7.2). We call *characteristic* a curve $t \longmapsto x(t)$ such that

$$\dot{x}(t) = f'(\rho(x(t), t).$$

The density ρ is constant along such curves, indeed:

$$\frac{d}{dt} \rho(x(t), t) = \partial_t \rho(x(t), t) + \dot{x}(t) \partial_x \rho(x(t), t)$$

$$= \partial_t \rho(x(t), t) + f'(\rho(x(t), t) \partial_x \rho(x(t), t) = 0.$$

Since ρ is constant along $t \longmapsto x(t)$, the speed is itself constant, and the characteristic is a straight line in the space-time domain:

$$t \longmapsto x(t) = x(0) + t f'(\rho_0(x(0))),$$

where ρ_0 is the density at time 0. If an initial density ρ_0 is given, it is then possible to build the associated solution by keeping the initial value constant along characteristics. The approach does make sense as far as characteristic curves *do not cross*.

For a given smooth initial density ρ_0, the flow associated to characteristic curves is

$$\Phi_t \; : \; x \longmapsto x + f'(\rho(x_0, 0))t.$$

If one assumes that f is C^2, one may compute the Jacobian of the transformation as

$$J(t, x) = 1 + t\, f''(\rho_0(x))\, \rho_0'(x).$$

This Jacobian remains positive (which means that the mapping is a diffeomorphism, i.e. characteristics do not cross) for all t if $f''(\rho_0(x))\, \rho_0'(x) \geq 0$. On the contrary, if the latter quantity is negative at some points, the mapping is regular only for

$$t < -\frac{1}{f''(\rho_0(x))\, \rho_0'(x)}.$$

The lifetime of the smooth solution is therefore

$$T = \frac{1}{\max |(f''(\rho_0(x))\, \rho_0'(x))_-|},$$

where $a_- > 0$ stands for the negative part of a (with the usual convention $a = a_+ - a_-$).

If one considers the standard flux $f(\rho) = U\rho(1 - \rho/\rho_{\max})$, we have that $f''(\rho) = -2U/\rho_{\max} < 0$. We shall therefore have a global smooth solution if ρ_0 is non-increasing. On the contrary, if ρ_0 is increasing, some characteristic curves will intersect as soon as t reaches the critical value $t_c = (f''(\rho_0(x))\, \rho_0'(x))^{-1}$, which rules out the possibility that a smooth solution exists beyond this time. This phenomenon necessitates to extend the notion of solution in order to give a general meaning to (7.2) for discontinuous functions. It leads to the notion of weak solution, which we briefly describe here: $\rho \in L^1_{\text{loc}}(\mathbb{R} \times]0, +\infty[)$ is called weak solution of (7.2) over $\mathbb{R} \times]0, +\infty[$ if $f(\rho) \in L^1_{\text{loc}}(\mathbb{R} \times]0, +\infty[)$ and if, for any C^1 function φ, compactly supported in $\mathbb{R} \times]0, +\infty[$, it holds that

$$\int_{\mathbb{R}} \int_0^{+\infty} \partial_t \varphi \, \rho(x,t) \, dx \, dt + \int_{\mathbb{R}} \int_0^{+\infty} \partial_x \varphi \, f(\rho(x,t)) \, dx \, dt = 0.$$

As an illustration, it can be checked that the characteristic function of an interval translated at constant speed v is a weak solution associated to the flux $f(\rho) = \rho v$. Such equations expressed in a weak form typically admit infinitely many solutions, and selecting the "right one", in order to establish well-posedness results, calls for the notion of entropy solution, which goes far beyond the scope of the present book. We refer the reader to the vast literature on this subject (e.g. Godlewski and Raviart, 1996).

Link with the microscopic model. A formal link with the microscopic model presented in Section 2.1 can be made, by noticing that the linear density (number of pedestrians per unit length) is the reciprocal of the distance between individuals: $\rho = 1/w$. If φ is the function which defines the [distance $\mapsto$ speed] mapping, it holds that

$$f(\rho) = \rho u(\rho) = \rho \varphi \left(\frac{1}{\rho} \right), \quad f'(\rho) = \varphi \left(\frac{1}{\rho} \right) - \frac{1}{\rho} \varphi' \left(\frac{1}{\rho} \right).$$

In the neighborhood of a stationary solution, with density ρ_{eq} and associated distance $w_{\text{eq}} = 1/\rho_{\text{eq}}$, we obtain

$$f'(\rho_{\text{eq}}) = \varphi(w_{\text{eq}}) - w_{\text{eq}} \varphi'(w_{\text{eq}}).$$

Equation (7.3) expresses a transport at speed $f'(\rho_{\text{eq}})$. The last term of the expression above corresponds to the propagation speed $-w_{\text{eq}} \varphi'(w_{\text{eq}})$ found in Section 2.1 at the microscopic level. The macroscopic speed includes an additional term, that is the speed of entities $\varphi(w_{\text{eq}})$. It comes from the fact that the macroscopic model is based on an *Eulerian* description (the macroscopic density is expressed in a fixed referential), in opposition to the microscopic approach, that is natively *Lagrangian* (the distances are attached to individuals in motion).

This approach has been extended in various directions, see e.g. Goatin and Scialanga (2016) for an account of the non-local dependence of the speed upon the density.

7.2. Two-Dimensional Models

As mentioned in the beginning of this chapter, the simplest macroscopic model is of the pure transport type: interactions are disregarded. In this setting, modeling aspects are restricted to the prescription of the velocity field. Consider e.g. a situation where all agents share a common goal Γ (like an exit or a set of exits) and have a full knowledge of the topography. It is then natural to assume that an agent starting from some position x shall tend to follow the shortest path to Γ. A path from an initial position x to the target Γ can be represented by a curve $\gamma : [0,1] \to \overline{\Omega}$ with $\gamma(0) = x$, $\gamma(1) \in \Gamma$. We denote by $\Lambda(x)$ the set of all those curves, and by $v(y)$ the speed at y. Now we consider that an individual located at x aims at minimizing their time to reach Γ, that is

$$\varphi(x) = \inf_{\gamma \in \Lambda} \int_0^1 \frac{|\gamma'(s)|}{v(x)}\, ds. \tag{7.4}$$

It can be established (see Section 8.2) that this time $\varphi(x)$, considered as a function of the initial position x, verifies the so-called non-homogeneous eikonal equation

$$|\nabla \varphi(x)| = \frac{1}{v(x)}. \tag{7.5}$$

The optimal strategy consists in heading to the direction $\nabla \varphi$, at the given speed $v(x)$, which yields a desired velocity defined as

$$U = -\frac{\nabla \varphi}{|\nabla \varphi|}\, v(x). \tag{7.6}$$

The formulation given in Hughes (2002) can be recovered by noticing that $U = -v(x)^2 \nabla \varphi$. Under the assumption that the local speed is the desired one $|U|$ which yields

$$\frac{\partial \rho}{\partial t} + \nabla \cdot (\rho U) = 0,$$

where U may also be written $-v(x)^2 \nabla \varphi$.

Congestion-induced speed reduction

The simplest way to account for congestion is to consider that the speed is affected by the local density. It actually consists in transposing LWR model (7.2) in the two-dimensional setting. Denote by U the desired velocity field (as previously defined). If we consider this field as given, the model simply consists in defining the actual velocity as the desired one multiplied by a correction factor to account for the local density $\beta = \beta(\rho)$:

$$\frac{\partial \rho}{\partial t} + \nabla \cdot (\rho \beta(\rho) U) = 0. \tag{7.7}$$

To account for the fact that people tend to reduce their velocity in crowded areas, it is natural to assume that β is non-increasing, with $\beta(0) = 1$, and $\beta(\rho) = 0$ as soon as ρ reaches a critical value $\rho_{\max}$. The simplest choice is

$$\beta(\rho) = 1 - \rho/\rho_{\max},$$

as proposed in Hughes (2002). Note that a similar expression is proposed in Dolak and Schmeiser (2005), under the name of *logistic sensitivity* in the context of Keller–Segel equations, to account for the inhibiting effect of congestion upon the chemotactic motion of cells.

Congestion-avoiding strategies

The approach proposed in Hughes' model (Hughes, 2002) incorporates more sophisticated features in terms of pedestrian behavior. In some way, it mixes both ingredients listed previously, namely instantaneous "computation" of the optimal path by each agent, and reduction of the speed in overcrowded areas. The idea consists in having the speed field v (used to define optimal paths) depends on the instantaneous density distribution. It can be defined as the product of the free speed v_0 by a correction factor $\beta(\rho)$. The potential is then defined as the solution to (7.5), the velocity is defined by (7.6), which yields (Hughes model):

$$\frac{\partial \rho}{\partial t} + \nabla \cdot (\rho f(\rho) \nabla \varphi) = 0 \tag{7.8}$$

with $f(\rho) = v_0^2 \, \beta(\rho)^2$.

Remark 7.1. The previous approach relies on the assumption that each agent has a full and instantaneous knowledge of the overall density distribution. Besides, since the crowd is in motion, strategies to avoid crowded areas based on the present configuration may turn out to be suboptimal. A more realistic approach consists in building individual strategies using local information only (see Carrillo *et al.*, 2016).

7.3.   Granular Models: Hard Congestion

Chapter 4 presents microscopic models based on a hard-sphere description of individuals, i.e. subject to a strict non-overlapping constraint. At the macroscopic level, such a description can be written by imposing the density to remain below a threshold value, which we shall set to 1 for simplicity.

Macroscopic granular model. The core ingredient, like in the microscopic case, is the field of desired velocity U. Since the description is Eulerian, we shall limit ourselves to the case of interchangeable persons in terms of behavior: the field $U = U(x)$ depends on the position only. In the case of the evacuation of a room, we shall identify the latter to a domain Ω of $\mathbb{R}^2$, this field points toward an exit, and thus tends to concentrate people. We assume that the total "mass" of people is 1, so that, for any t, ρ belongs to $\mathcal{P}(\Omega)$, the set of probability measures over Ω. The set of feasible densities is then (since we deal with measures which are absolutely continuous with respect to the Lebesgue measure, we shall identify the measure and the associated density):

$$K = \{\rho \in \mathcal{P}(\Omega),\ 0 \leq \rho \leq 1\ a.e.\}. \tag{7.9}$$

The macroscopic counterpart of the microscopic granular model of Chapter 4 is obtained by considering that the actual velocity of the crowd is the closest (in a L^2 sense) to the desired velocity U, among all feasible velocities. Informally, feasible velocities are such that the density should not be increased on zones where it already saturates the constraints, i.e. the divergence must be non-negative on the saturated zone (where $\rho = 1$). This property can be formalized in a dual way, by introducing the set of feasible velocities as

$$C_\rho = \left\{ v \in L^2(\Omega),\ \int_\Omega v \cdot \nabla p \leq 0 \quad \forall p \in H^1_\rho,\ p \geq 0\ a.e. \right\},$$

where the set H^1_ρ of pressure test functions is defined by

$$H^1_\rho = \{p \in H^1(\Omega),\ p(1 - \rho) = 0\ a.e.\}.$$

The macroscopic problem (macroscopic counterpart of (4.5)) simply writes

$$\begin{cases} \dfrac{\partial \rho}{\partial t} + \nabla \cdot (\rho u) = 0, \\[2mm] u = P_{C_\rho} U, \end{cases} \tag{7.10}$$

where the continuity equation is meant in a weak sense, and the projection on C_q corresponds to the L^2-norm.

Macroscopic granular model: mathematical issues. Since the mapping $\rho \mapsto u(\rho) = P_{C_\rho} U$ is non-smooth and non-local, this problem raises delicate issues from the theoretical standpoint. We shall restrict ourselves here to a short description of the adapted mathematical framework, and refer to Maury *et al.* (2011) for further details. The main idea consists in transposing the microscopic framework at the macroscopic level. The microscopic description is natively Lagrangian (the unknown $x_i(t)$, $u_i(t)$ are associated to a moving individual i), whereas the macroscopic one is Eulerian: $t \mapsto \rho_t(x)$ describes the evolution of the density at some point x which is fixed. Standard tools to analyze Partial Differential Equations are based on an Eulerian standpoint, in particular distances between functions are estimated by considering some integral norm of their difference, or the derivative of their difference. This Eulerian character rules out the possibility to adapt strategies like the Catching-up algorithm (see Eq. (4.16)). The framework of *Optimal transportation* (see e.g. Santambrogio (2015) or Villani (2003)) makes it possible to recover a partly lagrangian description of measures. This framework is based on the so-called Wasserstein distance, which can be defined on the space of probability measures $\mathcal{P}(\Omega)$. Let $\mathbf{t} : \Omega \longrightarrow \Omega$ be a measurable mapping, and $\mu \in \mathcal{P}(\Omega)$ a probability measure. We say that ν is the push-forward of μ by $\mathbf{t}$ if

$$\mu\left(\mathbf{t}^{-1}(A)\right) = \nu(A)$$

for any measurable set A. We then write $\nu = \mathbf{t}_\# \mu$. For given measures μ_0 and μ_1, we denote by $\Pi(\mu_0, \mu_1)$ the set of transport maps between μ_0 and μ_1. The quadratic Wasserstein distance $W_2(\mu_0, \mu_1)$ is defined by

$$W_2(\mu_0, \mu_1)^2 = \inf_{\mathbf{t} \in \Pi(\mu_0, \mu_1)} \int_\Omega |\mathbf{t}(x) - x|^2 \, d\mu_0(x).$$

Notice that this definition holds for atomless measures only, which is anyway the case that we are interested in.

We may now write down the Catching-up algorithm, or prediction–correction algorithm, as follows: given an initial density ρ^0 and a time step $\tau > 0$, build $\rho^1, \dots, \rho^k$ according to

$$\begin{cases} \tilde{\rho}^{k+1} = (\mathbf{id} + \tau U)_\# \, \rho^k & \text{transport (prediction)}, \\ \rho^{k+1} = P_K\left(\tilde{\rho}^{k+1}\right) & \text{projection (correction)}, \end{cases} \quad (7.11)$$

where the projection is performed in the Wasserstein sense and the set K is the constrained set introduced in (7.9). As detailed in Maury *et al.* (2010), it can be established that the projection on K for the Wasserstein distance, is properly defined in $\mathcal{P}(\Omega)$, and the sequence of discrete solutions can be shown to converge to a solution of the problem.

Remark 7.2. This well-posedness of the projection may appear paradoxical, in the light of Remark 4.5. Indeed, the macroscopic problem may be thought of (at least informally) as a limit of the microscopic model when the number of individuals goes to infinity, and their size goes to zero. Since the projection in the microscopic case degenerates in this asymptotic limit, a pathological behavior at the macroscopic could be expected. The fact that the projection turns out to be well defined in the macroscopic situation enlightens a deep difference between the many-body microscopic setting and the macroscopic ones, a difference which will have consequences in terms of modeling (see in particular the next paragraph on the Faster-is-Slower effect). This paradoxical effect is mainly due to the fact that the Wasserstein distance, while based on some Lagrangian description of nonnegative measures, is not fully Lagrangian. It is in particular insensitive to permutations of particles whereas, at the microscopic level, the Euclidean distance between position vectors in $\mathbb{R}^{2N}$ strongly depends on the indexing. We refer to Maury and Venel (2011) and Maury (2016) for further comments on this discrepancy between the two scales of description.

Model properties

We consider an evacuation situation like the one represented in Fig. 7.1. The saturated zone is denoted by ω, its boundary Γ consists of a wall part Γ_w, the exit Γ_{out}, and the upstream part Γ_{up} ("free" boundary of the saturated area). The saddle-point formulation of this projection problem takes the form of a unilateral Darcy problem

$$\begin{cases} u + \nabla p = U, \\ -\nabla \cdot u \leq 0 \end{cases} \tag{7.12}$$

in the saturated zone $\omega \subset \Omega$.

Neither Capacity Drop...

Assume that U is strictly concentrating, i.e. $\nabla \cdot U < 0$ in ω, the constraint is then activated over all the domain, so that $\nabla \cdot u = 0$. One may then

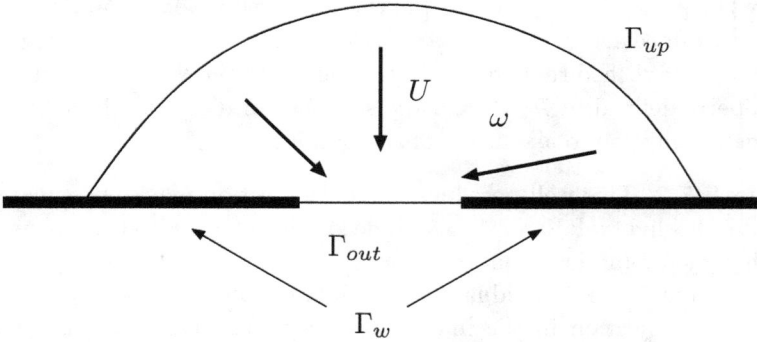

Fig. 7.1.   Macroscopic evacuation.

eliminate the velocity and obtain for p a Poisson problem

$$\begin{cases} -\Delta p = -\nabla \cdot \beta U & \text{in } \Omega, \\ p = 0 & \text{on} \Gamma_0 = \Gamma_{\text{out}} \cup \Gamma_{\text{up}}, \\ \dfrac{\partial p}{\partial n} = 0 & \text{on } \Gamma_w \end{cases} \qquad (7.13)$$

(the latter conditions simply express the impervious character of the wall: $u \cdot n = 0$).

Now, by the maximum principle, since the right-hand side is positive the pressure p is non-negative. As a consequence, it holds that $\partial p / \partial n \leq 0$, so that, on Γ_{out},

$$u \cdot n = U \cdot n - \nabla p \cdot n = U \cdot n - \frac{\partial p}{\partial n} \geq U \cdot n.$$

In other words, people on the boundary exit *faster* than they would if they were alone (i.e. walking at their desired velocity). This can be explained by recalling that the model has a mechanical nature. In particular, pressures implement interaction forces which obey the Law of Action–Reaction. In evacuation situations, it leads to people in the head being *pushed* from behind, thus accelerated. This model therefore exhibits some sort of anti-Capacity Drop phenomena (see Section 9.6).

... nor Faster-is-Slower effect

Let us now formalize and investigate the Faster-is-Slower issue, i.e.:

> *May a reduction of some individual desired velocities lead to a quicker evacuation?*

We consider again the situation represented by Fig. 7.1, with a desired velocity field U which points toward the exit. The question can be formalized by introducing a speed correction field $\beta(x)$, and the associated velocity field $\beta(x)\,U(x)$. The pressure p_β associated to this desired velocity is the solution to the Poisson problem

$$-\Delta p_\beta = -\nabla \cdot \beta U > 0,$$

with homogeneous Dirichlet boundary conditions $p_\beta = 0$ on Γ_{out} and Γ_{up}, and homogeneous Neuman boundary conditions $\partial p_\beta / \partial n = 0$ on Γ_w. The associated flux through the exit is thus

$$J(\beta) = \int_{\Gamma_{\text{out}}} u_\beta \cdot n = \int_{\Gamma_{\text{out}}} U \cdot n - \int_{\Gamma_{\text{out}}} \frac{\partial p_\beta}{\partial n}. \tag{7.14}$$

The question is then: considering the reference situation $\beta \equiv 1$, are there *positive* perturbations of β (i.e. people tend to go faster) which lead to a *decrease* of the flux J? The next proposition gives a precise answer to this question. We assume that desired velocities are not modified at the boundary Γ_{out}, i.e. $\beta = 1$.

Proposition 7.1. *Let us assume that the desired velocity U is tangent to the wall. Let $\beta \mapsto J(\beta)$ be the flux functional defined by (7.14). The gradient of J, that is the function g such that*

$$J(\beta + \tilde{\beta}) = J(\beta) + \int_\Omega g\tilde{\beta} + o(\tilde{\beta}),$$

for all variations $\tilde{\beta}$ vanishing on Γ_{out} is $U \cdot \nabla q$, where q is the solution to the adjoint problem

$$\begin{cases} -\Delta q = 0 & \text{in } \Omega, \\ q = 1 & \text{on } \Gamma_{\text{out}}, \\ q = 0 & \text{on } \Gamma_{\text{up}}, \\ \dfrac{\partial q}{\partial n} = 0 & \text{on } \Gamma_w. \end{cases} \tag{7.15}$$

Proof. Since the adjoint problem method is not so common in crowd motion modeling, we describe the approach with some details.

We first introduce a dual variable q, that is a function defined in Ω, and which makes it possible to express the relation between the control variable β and the state variable p. For any such q (assumed to be smooth), from (7.13), we have that

$$b(p,q) = \int_\Omega \nabla p \cdot \nabla q - \int_\Gamma \frac{\partial p}{\partial n} q + \int_\Omega q \nabla \cdot \beta U = 0, \qquad (7.16)$$

where the boundary integral over Γ can be replaced by an integral over $\Gamma_0 = \Gamma_{\text{out}} \cup \Gamma_{\text{up}}$, thanks to the boundary condition on Γ_w. Now, since we assume that desired velocities are not modified at the exit, and since we are interested in variations of the flux only, we may drop the first term of (7.14) in the definition of J:

$$J(\beta) = -\int_{\Gamma_{\text{out}}} \frac{\partial p_\beta}{\partial n}.$$

The core of the approach relies on a so-called *Lagrangian*, that is a function of the state variable p, the control variable β, and a dual variable q. It is defined as the sum of the objective function (flux) expressed in the state variable p (uncoupled from the control variable β), and the weak expression (7.16) of the state equation:

$$L(p,\beta,q) = \underbrace{-\int_{\Gamma_{\text{out}}} \frac{\partial p}{\partial n}}_{\text{flux}} + \underbrace{\int_\Omega \nabla p \cdot \nabla q - \int_{\Gamma_0} \frac{\partial p}{\partial n} q + \int_\Omega q \nabla \cdot \beta U}_{b(p,q)} = 0.$$

What is important here is that, for any p_β coupled with β by the state equation, $b(p_\beta, q) = 0$ for any q. As a consequence, for any β and any q,

$$L(p_\beta, \beta, q) = J(\beta).$$

We differentiate with respect to β both sides of this identity:

$$D_\beta J = D_p L \circ D_\beta p_\beta + D_\beta L. \qquad (7.17)$$

The approach consists in choosing an adjoint variable q such that $D_p L = 0$, which circumvents the difficulty to explicitly estimate $D_\beta p_\beta$. Let $\tilde{p}$ denote a variation in the variable p. It holds that

$$D_p L \, \tilde{p} = -\int_{\Gamma_{\text{out}}} \frac{\partial \tilde{p}}{\partial n} + \int_\Omega \nabla \tilde{p} \cdot \nabla q - \int_{\Gamma_{\text{out}}} \frac{\partial \tilde{p}}{\partial n} q - \int_{\Gamma_{\text{up}}} \frac{\partial \tilde{p}}{\partial n} q$$

$$= -\int_{\Gamma_{\text{out}}} \frac{\partial \tilde{p}}{\partial n}(q+1) + \int_\Omega (-\Delta q)\tilde{p} + \int_\Gamma \frac{\partial q}{\partial n}\tilde{p} - \int_{\Gamma_{\text{up}}} \frac{\partial \tilde{p}}{\partial n} q.$$

Since $\tilde{p}$ is a variation of p, which identically vanishes on Γ_0, the boundary integral over $\Gamma = \partial\Omega$ reduces to an integral over Γ_w. The adjoint problem is designed in order to vanish the previous expression:

$$-\Delta q = 0 \text{ in } \Omega, \quad q = -1 \text{ on } \Gamma_{\text{out}}, \quad \frac{\partial q}{\partial n} = 0 \text{ on } \Gamma_w, \quad q = 0 \text{ on } \Gamma_{\text{up}},$$

which is exactly (7.15), up to a change in sign in the Dirichlet boundary condition. Let q be the solution of this adjoint problem, so that $D_p L(p_\beta, \beta, q) = 0$. From (7.17), it comes

$$D_\beta J = 0 + D_\beta L(p_\beta, \beta, q),$$

with

$$D_\beta J \tilde{\beta} = \int_\Omega q \nabla \cdot \tilde{\beta} U = -\int_\Omega \tilde{\beta} U \cdot \nabla q + \int_\Gamma \tilde{\beta} q U \cdot n.$$

Since $\tilde{\beta} = 0$ on Γ_{out}, $q = 0$ on Γ_{up}, and $U \cdot n = 0$ on Γ_w, the boundary term vanishes, which yields

$$\nabla_\beta J = -U \cdot \nabla q.$$

If we denote by q the solution to (7.15), that is the opposite of the solution to the native adjoint problem, we obtain the announced expression (without the minus sign). □

The previous proposition gives a precise answer to the Faster-is-Slower issue. Let q be the solution to the adjoint problem (7.15). As soon as

$$U \cdot \nabla q \geq 0 \tag{7.18}$$

overall the congested zone, no FiS effect can be reproduced in this framework. More precisely, any local or global increase of the speed (i.e. $\beta = 1 + \tilde{\beta}$, with $\tilde{\beta}(x) \geq 0$ for every x) leads to an *increase* of the flux, simply because

$$\nabla_\beta J \cdot \tilde{\beta} = \int_\Omega \tilde{\beta} U \cdot \nabla q \geq 0.$$

It is also important to notice that the condition $U \cdot \nabla q \geq 0$ can be expected to be verified in general, for any reasonable way to define the desired velocity upstream an evacuation exit.

7.4. Micro–Macro Issues

In the one-dimensional setting, the LWR model (7.2) presented in the beginning of this chapter can be obtained from the microscopic FTL model (see Chapter 2). In Colombo and Rossi (2014), the authors investigate the convergence of the FTL (microscopic) model

$$\frac{dx_i}{dt} = \varphi(x_{i+1} - x_i), \quad 1 \le i \le N,$$

with $x_{N+1}(t)$ prescribed, toward the LWR (macroscopic) model

$$\partial_t \rho + \partial_x(\rho u(\rho)) = 0.$$

The approach relies on an operator which maps a microscopic configuration to a macroscopic density. More precisely, if one considers the entity i as a segment of unit density and length w_m, centered at x_i, half the mass of i and $i + 1$ is uniformly spread out between x_i and x_{i+1}, which scales it like $w_m/(x_{i+1} - x_i)$. More formally, the macroscopic density associated to a configuration $x = x_1, \ldots, x_N)$ is

$$\rho_x = \sum_{i=1}^{N} \frac{w_m}{x_{i+1} - x_i} \mathbf{1}_{[x_i, x_{i+1}[},$$

as illustrated by Fig. 7.2. The reciprocal of this operator can be built in a straightforward way by leftward integrating mass from the right end, and positioning an entity center each time a multiple of w_m is attained. The correspondence between the microscopic behavioral function $w \mapsto \varphi(w)$ and its macroscopic counterpart $\rho \mapsto u(\rho)$ is prescribed accordingly:

$$\varphi(w) = u\left(\frac{w_m}{w}\right) \quad \text{or,} \quad \text{equivalently,} \quad u(\rho) = \varphi\left(\frac{w_m}{\rho}\right).$$

The convergence result essentially states that, given an initial density, it can be transformed into a microscopic configuration with n entities (so that an entity scales like $1/n$). The associated microscopic solution can be mapped

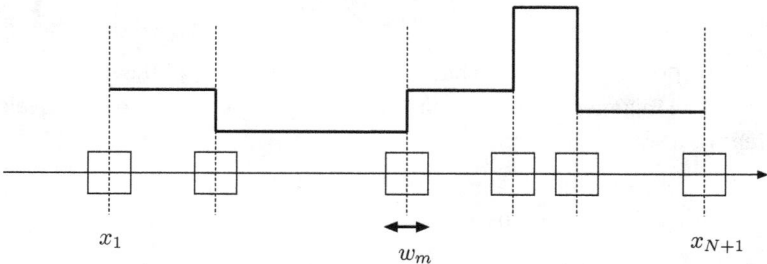

Fig. 7.2.   Micro–macro mapping.

to a path of diffuse densities. The latter can be shown to converge, when n goes to infinity, to the solution to the macroscopic problem. We also refer to Cristiani and Sahu (2016) for an extension of this micro–macro limit for a network, and to Goatin and Rossi (2017) for a similar convergence result (expressed in Wasserstein distance) in the case of a transport equation with a non-local term (the local velocity depends on the density in a certain neighborhood, through a integration kernel).

Note that the hard-sphere setting, which highly constrains relative motions of disks, is lost at the macroscopic scale. The macroscopic constraints simply expresses, at each position, a global limitation of density increases, whereas the non-overlapping constraints rules out motions of disks in various directions (see Maury *et al.* (2011) or Maury (2016) for further details on those micro–macro discrepancies).

7.5. Alternative Macroscopic Models

A huge number of macroscopic approaches have been proposed to model crowd motions, most of which go far beyond the scope of the present book. We give a short description of some of those approaches. In Degond *et al.* (2017), the authors propose a model based on the Euler equation, adapted to handle crowd hard congestion with a non-uniform bound on the local density. In Degond *et al.* (2013), macroscopic models of the kinetic types are derived from microscopic interaction rules which encode individual anticipation (collision avoidance).

Let us also mention a comparison of macroscopic models of the Hughes' type proposed in Twarogowska (2014), in particular in the context of room evacuation, and also a book (Bellomo *et al.*, 2017) which gives a wide description of general issues in the mathematical modeling of living entities.

Chapter 8

Computing Distances and Desired Velocities

As pointed out in the introduction, the key notion in crowd motion modeling is that of *desired velocity*. In some situations, e.g. when pedestrians are in a row (like drivers in a one-way road), the problem is reduced to that of defining a desired *speed* for each individual. In the most general situation, e.g. large crowds strolling in a mall or in an exhibition center, such a task is obviously out of reach. Inferring the desired velocities of individuals would require a knowledge of their instantaneous state of mind, personal history, intimate tastes. We shall therefore limit ourselves to standard, and somewhat idealized, situations, where individuals (or at least large classes of individuals) share a common goal.

The typical situation which we shall consider here corresponds to the evacuation of a building. The geometric model of a building consists in defining a domain of the plane, delimited by a boundary Γ, decomposed in an impervious part Γ_w (walls, border of obstacles, ...), and a part Γ_{out} open to the outside world, possibly made of several connected components (multiple exits). The core notion is that of *geodesic distance* to the outlet. Given a point x in the domain, this distance $D(x)$ is the length of the shortest path to Γ_{out}.

In the case of a convex room with a single exit, or a small number of exits, computing this distance is straightforward. Figure 8.1 presents shortest paths in a convex room with a single exit (top-left), and with two exits (top-right). When the room is no longer convex, e.g. because of

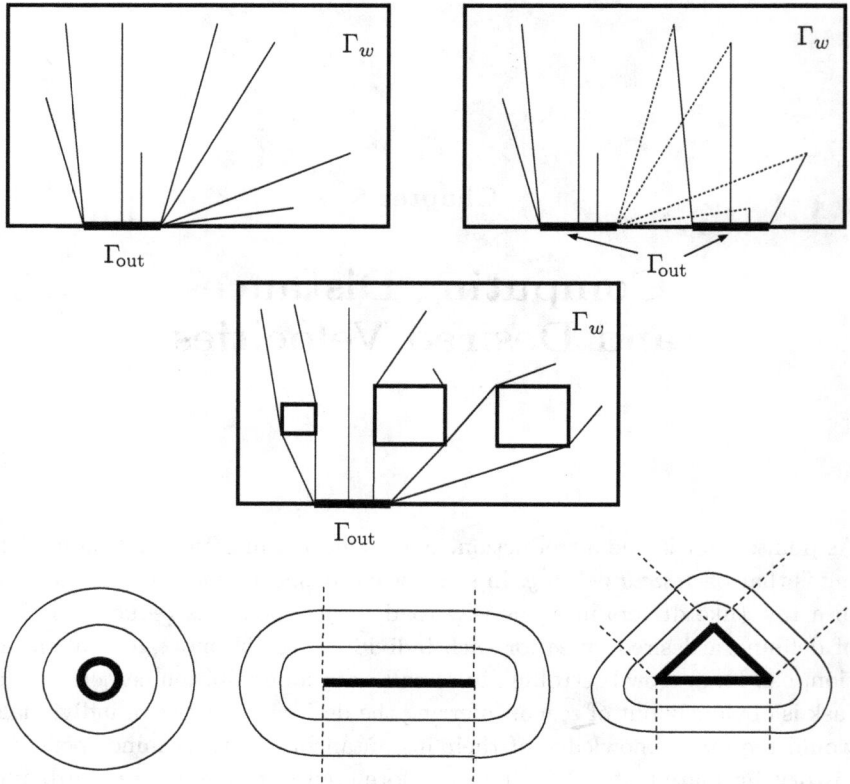

Fig. 8.1.   Shortest paths, and distances to obstacles.

obstacles, the geodesic distance (together with shortest paths) can still be analytically computed in simple situations (like in Fig. 8.1, middle).

From geodesic distance to desired velocities. Once the geodesic distance $D(x)$ to an exit or a set of exits has been computed, the desired velocity can be defined by simply considering that each individual walks in the steepest descent direction with respect to this distance. If we denote by v the speed of the considered individual, the desired velocity is defined by

$$U = -\frac{\nabla D}{|\nabla D|} v. \tag{8.1}$$

Remark 8.1. Note that the speed may vary from one individual to another. One may also consider that the speed, for a given individual, is not uniform, i.e. $v = v(x)$. This extension is to be considered with precaution: if the speed is not uniform, then the gradient of the geodesic distance might not be relevant here, but rather the shortest path based on the overall distribution of speeds. For instance, if a zone is locally impaired, with a much smaller local speed, the associated shortest path is likely to circumvent this zone, changing the direction of the desired velocity, and not only its speed. This standpoint is considered in this chapter (see Eq. (8.7)), and it calls for non-trivial computations. We note, however, that building the direction on the straight geodesic distance while considering local speed changes may make sense in some situations. Consider for example individuals evacuating a building with a good knowledge of the topography, thereby having some sort of inner local map to the exit. If such individuals meet unexpected problems (like smoke, obstacle on the floor), they may have to reduce their speed without having the possibility to instantaneously "re-compute" an optimal path based on the new distribution of speed, simply because they do not have a full knowledge of the overall situation (neither present, nor future).

Distance to obstacles. Handling of obstacle constraints also requires an estimation of the distance to them. Again, as far as the shape of obstacle is standard, distances can be straightforwardly computed. Isovalues of the distance function to some simple geometric domains are represented in Fig. 8.1, bottom (disk, segment, triangle).

Yet, in the general situation of a irregular room, with many obstacles of general shapes, such an approach is not tractable. Furthermore, it is sometimes relevant to consider that the typical speed of a pedestrian may depend on various factors, like the local visibility (e.g. presence of smokes), the quality of the floor, or the local density of the crowd. An efficient computation of the geodesic distance therefore calls for more general and sophisticated tools, which are the core of the present chapter. We shall mainly address *shortest path* problems in the context of Dynamic Programming. To emphasize the general character of the underlying principles, the first section (8.1) is focused on shortest path problem set on graphs. The approach is then instantiated in the case of Euclidean domains (like rooms or buildings).

8.1.   Shortest Path Problem on a Graph

We first consider the problem on a graph, this will emphasize the role played
by Dynamic Programming principles in setting the problem in a global way,
and to numerically solve it in an efficient way. Besides, the approach which
we present here is directly applicable to the search of optimal path in a
complex building, set of buildings, or even large scale complexes. From this
standpoint, edges will correspond to single paths, corridors, stairways, or
even metro or bus trips ..., and vertices to bifurcation points, entrances,
exits, ... In order to anticipate the application to Euclidean domains, we
shall consider that edges are characterized by a static property, that is the
length, and by a possibly dynamic property, that is the mean velocity of
an individual going along this edge. We assume here that this velocity is
constant.

We thus consider an undirected graph $\mathcal{G} = (V, E, \ell, v, \Gamma)$, where V is a
(finite) set of vertices, $E \subset V \times V$ the set of edges (assumed to be symmetric,
i.e. $(x, y) \in E \Leftrightarrow (y, x) \in E$), $\ell(\cdot)$ a collection of positive lengths defined on
edges, and $v(\cdot)$ a collection of nonnegative mean speeds, defined on edges
also. We single out of V a target set Γ, and we consider the problem which
consists in reaching Γ, starting from any vertex $x \in V$. We denote by $\mathcal{C}$ the
set of finite chains on the graph:

$$\mathcal{C} = \{\xi = (x_0, \ldots, x_{n_\xi}) \in V^{n_\xi+1}, \ n_\xi \in \mathbb{N}, \ (x_{i-1}, x_i) \in E, \ \forall i \leq n_\xi\},$$

and by $\mathcal{C}(x, \Gamma)$ the subset of chains starting at x and ending in Γ (i.e. $x_0 = x$
and $x_{n_\xi} \in \Gamma$). We denote by $\varphi : V \longrightarrow \mathbb{R}_+$ the so-called *value function*
defined by

$$\varphi(x) = \min_{\xi \in \mathcal{C}(x,\Gamma)} \sum_{i=0}^{n_\xi-1} \frac{\ell(x_i, x_{i+1})}{v(x_i, x_{i+1})}, \tag{8.2}$$

that is the shortest possible time to reach Γ, starting from x. The fact that
the infimum is attained is obvious, since the number of chains below a given
length is finite.

The core of the approach relies on the *Bellman equation*, which simply states a local optimality principle: starting from some vertex y and
considering the finite set of available choices, each choice $x \sim y$ yields a
value that is the cost of the single step (y to x), plus the shortest path
starting from x. The value at y obviously corresponds to the best one-step

strategy:

$$\varphi(y) = \min_{x \sim y} \left(\frac{\ell(x,y)}{v(x,y)} + \varphi(x) \right). \tag{8.3}$$

Note that, since $\varphi(y) > \varphi(x_{\min})$, where $x_{\min}$ realizes the minimum, this minimum can be restricted to all those $x \sim y$ such that $\varphi(x) < \varphi(y)$.

This principle can be implemented as the elementary brick of an algorithm to compute all the values $\varphi(y)$, by *upstream propagation of the information from the target*. The algorithm reads as follows.

Algorithm 8.1. We consider a graph $\mathcal{G} = (V, E, \ell, v, \Gamma)$. The value function φ is defined by (8.2), i.e. $\varphi(x)$ is the minimal time needed to reach the target $\Gamma \subset V$, starting from $x \in V$. We denote by X the subset of V on which the values $\varphi(x)$ have already been computed, that is the so-called *frozen* zone. We shall keep the notation X to designate this evolving set.

Initialization.
The algorithm is initialized by taking $X = \Gamma$, $\varphi \equiv 0$ on X.

Front propagation.
We consider that φ has been computed on X. We define B (*narrow band*) as the set of vertices outside of X that are connected to some vertex in X. We predict the values of φ at all vertices in B:

$$\tilde{\varphi}(y) = \min_{x \sim y,\, x \in X} \left(\frac{\ell(x,y)}{v(x,y)} + \varphi(x) \right). \tag{8.4}$$

We define $y_{\min} \in B$ as a vertex which minimizes φ over B. This vertex is added in X, and the value $\varphi(y_{\min})$ is set to $\tilde{\varphi}(y_{\min})$.

The propagation step is repeated until X covers V.

Note that, when the algorithm enters the propagation step, most values $\tilde{\varphi}(y)$ have already been computed. The task simply consists in updating $\tilde{\varphi}$ at neighbors of $y_{\min}$ which were already in B (the minimum in (8.4) is computed over a set that has been enriched), adding new vertices to B (neighbors of $y_{\min}$ which were neither in B, nor in X), and compute $\tilde{\varphi}$ from (8.4) at those points. The critical task in terms of efficiency is the determination of the smallest distance at each step, and the efficiency of the overall algorithm heavily relies on both the data structure and the sorting strategy.

Remark 8.2. The so-called *heap* structure (see e.g. William, 1964) is particularly adapted to this situation. It consists in keeping an account of vertices of B within a tree structure which respects the ordering associated to the current predicted distance $\tilde{\varphi}(y)$ (called the *key* in this context). More precisely, the distance of any vertex is smaller than the distances associated to all their descendants. The heap is modified at two occurrences in each step:

(1) *Root removal.* According to the heap structure, $y_{\min}$ is the root. When it is transferred in X, it disappears and the tree is restructured in order to preserve the heap property.
(2) *Incoming of new vertices.* When a vertex is added to B, it is added as a leaf to one of the shortest branches (or it starts a new generation if the tree is complete), and is then moved to its right position by performing binary permutations along the branch that it belongs to.

8.2. Shortest Path on a Domain: The Eikonal Equation

We consider a domain Ω in $\mathbb{R}^2$ (that covers the room or the building in which pedestrians evolve) and a subset Γ of $\overline{\Omega}$, that is the *target set*, it typically corresponds to the exits in case of an evacuation. We furthermore introduce a (scalar) speed field $v(x)$ defined on Ω, which corresponds to the local speed of individuals. The fact that v is not uniform makes it possible to implement geometric non-uniformities. For instance, the proximity to an obstacle, or a reduced visibility, may locally reduce the walking speed. For any $x \in \Omega$, we denote by $\varphi(x)$ the shortest time needed to reach Γ starting from x, given the speed field $v(\cdot)$. In other words, if we denote by Λ the set of curves $\gamma : [0,1] \to \overline{\Omega}$ with $\gamma(0) = x$, $\gamma(1) \in \Gamma$, we define

$$\varphi(x) = \inf_{\gamma \in \Lambda} \int_0^1 \frac{|\gamma'(s)|}{v(x)}\, ds, \qquad (8.5)$$

which is the straight continuous counterpart of (8.2).

We shall now establish in an informal way that φ verifies the so-called *Eikonal Equation*. Consider $x \in \Omega$, and $\varepsilon > 0$ a time increment. A local strategy consists in choosing a direction n (unit vector), walk at speed $v(x)$ in this direction during ε, and then take the shortest path from $x + v(x)\varepsilon n$ to Γ (see Fig. 8.2). The total time associated to this strategy n is then $\varepsilon + \varphi(x + v(x)n\varepsilon)$. We thus have (the sign "$\approx$" accounts for the fact that the speed during the ε-hop is not exactly constant) the Bellman-like equation

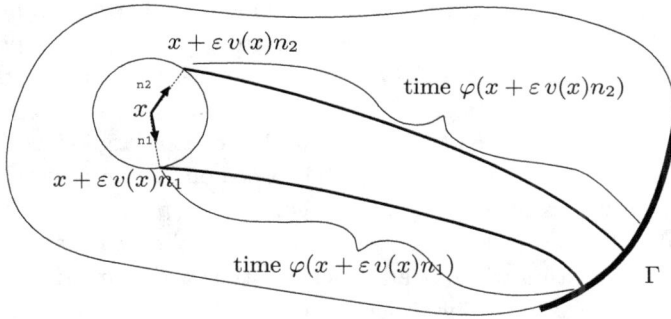

Fig. 8.2.   Shortest paths.

(continuous counterpart of (8.3))

$$\varphi(x) \approx \inf_{|n|=1} \left(\varepsilon + \varphi(x + \varepsilon\, v(x)\, n) \right). \tag{8.6}$$

Minimality requires n to be chosen along the steepest descent direction of φ, i.e. $n = -\nabla\varphi / |\nabla\varphi|$, which yields

$$\varphi(x) \approx \varepsilon + \varphi\left(x - \varepsilon\, v(x) \frac{\nabla\varphi}{|\nabla\varphi|} \right) = \varepsilon + \varphi(x) - \varepsilon\, v(x)\nabla\varphi \cdot \frac{\nabla\varphi}{|\nabla\varphi|} + o(\varepsilon)$$

$$= \varphi(x) + \varepsilon\left(1 - |\nabla\varphi|\, v(x) \right) + o(\varepsilon).$$

We finally obtain the eikonal equation

$$|\nabla\varphi(x)| = \frac{1}{v(x)}, \tag{8.7}$$

with $\varphi \equiv 0$ on the target Γ.

Space discretization. We shall describe here a basic scheme to solve Eq. (8.7) in a Euclidean domain, in the homogeneous case (v is taken constant, equal to 1), and we refer to Sethian (1999) for extensions to more general cases, and further details on implementational issues. The strategy relies on two main ingredients: (1) a front propagating approach similar to the fast marching algorithm on a graph (Algorithm 8.1), (2) a space discretization strategy for Eq. (8.7). Since we consider the homogeneous situation, the value function in which we are interested can be taken equal to the geodesic distance (which is proportional to the time in the homogeneous situation), and we shall consequently denote it by D. The first step consists in covering the domain of interest by a cartesian grid, the cells of which are indexed by couples (i,j) (canonic two-dimensional cartesian indexing). The algorithm consists in defining explicit rules to compute approximate values of D at the centers of the $h \times h$ square cells. We initialize the value of D at the target cells (e.g. cells which correspond to the targeted exit) to 0, and we set the value at obstacles, wall, etc. at a large value.

Like in Algorithm 8.1, the approach is based on growing a set X of cells in which the values have already been computed (*frozen* set). This set is initialized as the target set. Each step of the algorithm reads as follows (we present the algorithm without any consideration with numerical efficiency in terms of complexity, and we refer to Remak 8.3 below for some additional comments on those aspects):

Propagation step. We consider that D has already been computed on the cells of the frozen set X. We define B (*narrow band*) as the set of cells outside of X that share an edge with some cell in X (see Fig. 8.3). We compute predicted values for D in each cell of the narrow band B according to the following procedure: let (i,j) be one of those. We introduce

$$a = \min(D_{i-1,j}, D_{i+1,j}) \quad \text{and} \quad b = \min(D_{i,j-1}, D_{i,j+1}).$$

In the previous expressions $D_{k,\ell}$ is taken equal to its computed value if (k,ℓ) belongs to the current frozen zone, and to $+\infty$ (a large value in practice)

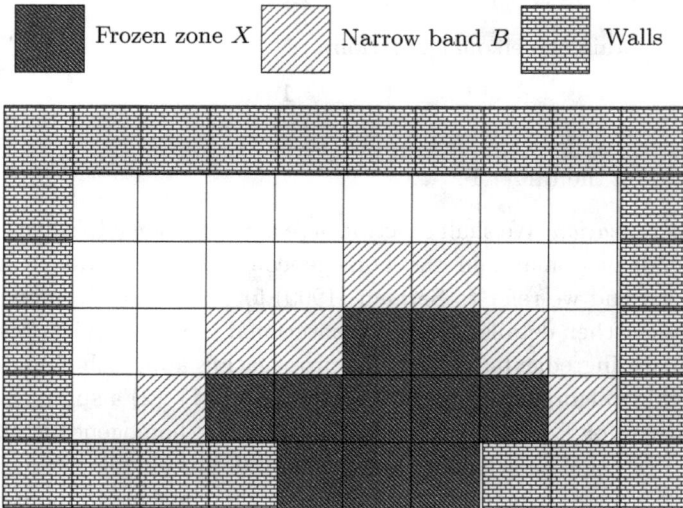

Fig. 8.3.   Front propagation (fast marching algorithm).

otherwise (i.e. for (k, ℓ) in the narrow band, or in obstacles/walls). We then set

$$
\tilde{D}_{i,j} = \begin{cases} \dfrac{a + b + \sqrt{2h^2 - (a-b)^2}}{2} & \text{if } |a - b| < h, \\[2ex] h + \min(a, b) & \text{if } |a - b| \geq h. \end{cases}
$$

Once all predicted distances $\tilde{D}$ have been computed in the narrow band, one picks up a cell (k, ℓ) which realizes the minimal value in B, this cell is added to the frozen zone X with its associated value $D_{k,\ell} = \tilde{D}_{k,\ell}$, and we proceed to the next propagation step, until X covers the whole domain.

Remark 8.3 (Implementational issue). A straight programming of the previous procedure would lead to a prohibitive cost. Like in the graph situation (Algorithm 8.1), a considerable reduction of this cost can be achieved as follows. First, when entering a propagation step, most predicted distances have already been computed, only the values in cells that are adjacent to (k, ℓ) (the previous "winner") need an update. Besides, a *heap* structure can be used in the present situation to keep track of a partial ordering of the cells in the narrow band (see Remark 8.2), with respect to the predicted distances. It makes it possible to rapidly identify a cell which realizes the minimum, without performing a full sorting over the narrow band cells.

Remark 8.4 (Consistence of the space discretization scheme). If we define

$$
\Delta_{ij}^{-x} = \frac{D_{i,j} - D_{i-1,j}}{h}, \quad \Delta_{ij}^{+x} = \frac{D_{i+1,j} - D_{i,j}}{h},
$$

and $\Delta_{ij}^{-y}, \Delta_{ij}^{+y}$, in a similar way, it can be checked that

$$
\max(\Delta_{ij}^{-x}, -\Delta_{ij}^{+x}, 0)^2 + \max(\Delta_{ij}^{-y}, -\Delta_{ij}^{+y}, 0)^2 = 1,
$$

which is a space discretized version of the Eikonal equation (8.7) (for $v \equiv 1$).

8.3. Non-homogenous Domains, Various Extensions

This section addresses some extensions of the previous approach. In particular, it may be relevant to have the local speed (denoted by $v(x)$ in

Eq. (8.7)) depend on the relative position to obstacle (e.g. it is not comfortable to walk at full speed in the very neighborhood of a wall). We also investigate some ways to provide smoother, and therefore more realistic, optimal trajectories.

Obstacles, static heterogeneities

We denote by U the "free" speed. One can account for the tendency of pedestrians to avoid the very neighborhood of obstacles by reducing the speed therein. If we denote by $D_o(x)$ the distance of a point x to the set of obstacles, one can define

$$
v(x) = \begin{cases} U & \text{if } D_c \leq D_o(x), \\ U\dfrac{D_o(x)}{D_c}\left(1 - \dfrac{D_o(x)}{D_c}\right) & \text{if } 0 \leq D_o(x) \leq D_c, \end{cases} \tag{8.8}
$$

where D_c is the range of influence of the obstacles, typically of the order of people diameter.

Figure 8.4 represents some computed shortest paths for the homogeneous Eikonal equation (top left), and for the modified Eikonal equation with non-homogeneous speed field $V(x)$ computed according to (8.8) (top right). This approach requires the knowledge of the distance to the set of obstacles. The distance D_o is actually pre-computed in a similar manner, by solving an eikonal equation.

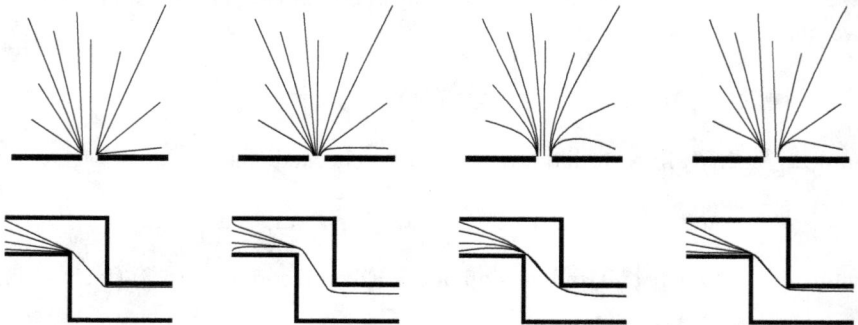

Fig. 8.4. Computed optimal paths (from top to bottom): homogeneous Eikonal equation, non-homogeneous case (repulsive effect of walls), Elastica model, Dubins' model (courtesy of C. Berenfeld and J.M. Mirebeau).

More sophisticated tools can be used to increase the realism of the computed shortest paths. In particular, pedestrians are reluctant to abruptly change their direction. It can be accounted for by changing the functional to minimize. In the basic version presented above, individuals tend to minimize their own evacuation time, possibly accounting for the fact that the local speed may vary (because of obstacles, or congested zones). A new approach has been proposed recently (Mirebeau, 2017). It is based on a local metric which depends on the curvature. The underlying theoretical framework goes far beyond the scope of this book, but the associated software Hamilton-FastMarching (Mirebeau, 2017) can be used to compute "optimal paths" in this new sense: they minimize the time to reach the prescribed target (the exit), plus an extra term which penalizes the local curvature of the path. We illustrate this approach in Fig. 8.4. We plot the computed paths in two cases: the so-called *Elastica* model, which penalizes the integral of the square curvature, and the *Dunbins'* model, which forces the curvature to remain below a prescribed threshold value.

Accounting for congestion

We consider here that a pedestrian has a fair knowledge of the global density distribution, this knowledge can be integrated in the computation of shortest (in the time sense) paths by considering the non-homogeneous Eikonal equation (8.7) with v defined as a decreasing function of the local density ρ, e.g.

$$v = U \left(1 + \omega\rho\right)^{-1}$$

as proposed in Hartmann *et al.* (2014) in the context of cellular automata, or in a more standard way $v = U \left(1 - \rho/\rho_{\max}\right)$. This very approach is the core of the celebrated Hughes' model (Hughes, 2002), based on a macroscopic description of the crowd. It is referred to as *adaptive route choice* model in Duives *et al.* (2013). In the context of microscopic models, this approach necessitates to compute in a relevant manner a diffuse density from the discrete configuration of people. We illustrate here the approach by prescribing a diffuse density which tends to slow down pedestrians. Figure 8.5 presents optimal paths in two such situations. On the left-hand side of the figure, we prescribe localized high density zone (the closed contours represent the isovalues of this density), to model small static groups of people.

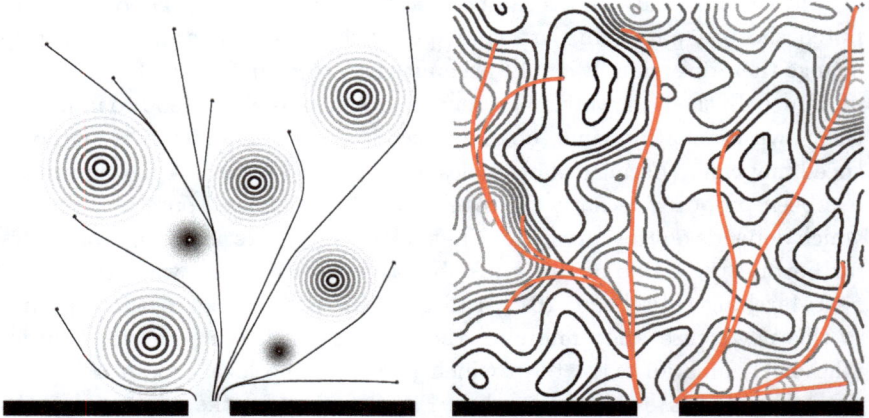

Fig. 8.5.   Optimal paths in a non-homogeneous environment.

The plot on the right represents a general situation, with a complex density landscape which tends to deflect trajectories. Note that the corresponding paths are straight analogs of light rays in a non-homogeneous environment (variable refractive index).

Remark 8.5 (Shortest paths vs. gradient flows). Let us underline the deep difference between this computation of optimal paths and the seemingly similar framework of gradient flows. To start with, consider the situation presented in Fig. 8.5 (left). This situation could be modeled by defining a potential as the sum of the distance to the exit and repulsive penalty terms accounting for dense areas, in the spirit of Fig. 3.10. The steepest descent paths are likely to look similar to the ones obtained by resolution of the non-homogeneous eikonal equation. Yet, both settings are very different. To lighten this difference, consider a congested area with a non-convex shape, like the banana-like zone presented in Fig. 8.7. In a gradient flow setting, the trajectories are based on local information only: a person starting from the middle of the room would be stuck upstream the exit in a local minimum of the potential. In the present setting, optimal path computations are based on a global knowledge of the density distribution, in such a way that the computed paths avoid the obstacle.

8.4. Shortest Paths in a Dynamic Environment

We previously considered the local speed $v(x)$ as independent from the time. In the situation where v is made dependent on the local density of individuals, it corresponds to the motion of a single individual in a *frozen* crowd, which obviously makes little sense. Besides, it may be of interest to handle a dynamic environment, for instance an access door which is closed at a certain time (assuming that all participants know it). In this spirit, consider the situation where v explicitly[1] depends on the time also, $v = v(x,t)$. The approach followed for the stationary case can be carried out in the present setting. We denote by $\varphi(x,t)$ the time associated to the shortest path to Γ for an individual starting at x at time t. The counterpart of Eq. (8.6) is now

$$\varphi(x,t) \approx \inf_{n} \left(\varepsilon + \varphi(x + \varepsilon\, v(x)\, n, t + \varepsilon) \right),$$

which leads to

$$\frac{\partial \varphi}{\partial t} - |\nabla \varphi|\, v(x,t) = -1.$$

From a computational standpoint, this approach requires a full space-time computation of φ, since φ is only known on Γ (it is identically 0). Even if one is interested on the optimal path of a single individual only, the full space-time computation seems unavoidable, since it is impossible to determine *a priori* at what time an individual starting at x will hit the target.

Optimal path for a single individual in a moving environment. The optimal strategy for a single starting point can be obtained by a straight fast marching algorithm, by simply reversing the standpoint, i.e. by propagating information from the starting point, rather than from the target. Consider an individual at $x \in \Omega$, aiming at reaching Γ, in a moving environment encoded by $v(x,t)$. Without loss of generality, we consider that the individual starts from x_0 at time 0 (see Fig. 8.6). We now denote by $\varphi(x)$ *the time taken by this individual to reach* $x \in \Omega$. The counterpart of

---

[1]Let us emphasize that, in the case where v is made dependent on the local density, the approach remains questionable, since the density *in the future* depends on past strategies, which themselves depend on the future. We refer to the *critical discussion* below for some remarks on those issues, which go far beyond the scope of this book.

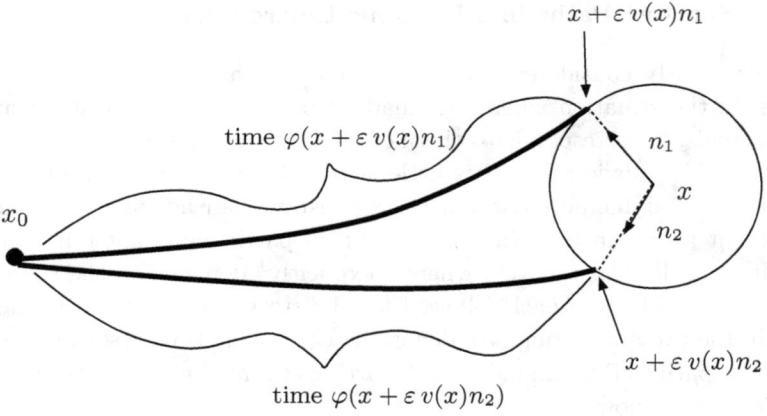

$x + \varepsilon\, v(x)n_1$

time $\varphi(x + \varepsilon\, v(x)n_1)$

n_1

x_0

x

n_2

$x + \varepsilon\, v(x)n_2$

time $\varphi(x + \varepsilon\, v(x)n_2)$

Fig. 8.6.   Shortest paths in a moving environment.

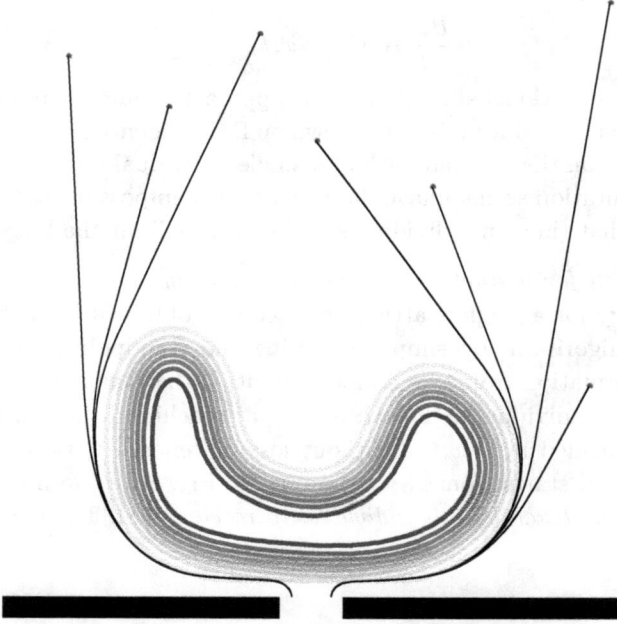

Fig. 8.7.   Banana test case (courtesy of C. Berenfeld and J.M. Mirebeau).

Eq. (8.6) is now

$$\varphi(x) \approx \inf_{|n|=1} (\varepsilon + \varphi(x + \varepsilon\, v(x, \varphi(x))\, n))$$

$$\approx \varepsilon + \varphi\left(x - \varepsilon\, v(x, \varphi(x))\frac{\nabla\varphi}{|\nabla\varphi|}\right)$$

$$\approx \varepsilon + \varphi(x) - \varepsilon\, v(x, \varphi(x))\nabla\varphi \cdot \frac{\nabla\varphi}{|\nabla\varphi|}$$

$$= \varepsilon + \varphi(x) - \varepsilon\, v(x, \varphi(x))\, |\nabla\varphi|\,,$$

which leads to

$$|\nabla\varphi(x)| = \frac{1}{v(x, \varphi(x))}. \qquad (8.9)$$

The time to reach Γ is then

$$T_\Gamma = \inf_{x\in\Gamma} \varphi(x) = \varphi(x_\Gamma)$$

and the optimal path is the streamline of $-\nabla\varphi$ which joins x_Γ to x_0.

In spite of the highly nonlinear character of Eq. (8.9), it can be straight-forwardly solved by a fast marching algorithm, *starting from the source x_0* (while the arrival time is not known, the departure time is).

Critical discussion. The approach presented previously is based on the assumption that each pedestrian has a full knowledge of the overall crowd, which may be ruled out in practice: as soon as the room is not star-shaped with respect to their current location, direct vision is impossible. Besides, even in a convex room, average-sized individuals in a dense crowd have a perception of others that is limited to their very neighborhood. From this standpoint, our virtual pedestrians are idealized, in the sense that they are unrealistically more efficient than the real ones. But they are also *less* efficient from another standpoint, since optimal path computation is based on the *current* configuration, disregarding any anticipation process. Accounting for this anticipation highly increases the complexity of the problem, since the future density field depends on previous strategies, which themselves depend on future densities. This problematic is the core of a wide field of research, which extends far beyond crowd motion, and far beyond the scope of this book, called Mean Field Games. For further details, we refer to the seminal paper (Lasry and Lions, 2007), and to a more recent contribution with applications to crowd motions (Benamou *et al.*, 2017).

From Optimal Times to Desired Velocities

We previously presented a general strategy to solve the general problem: given a target set Γ, and a speed field $v(\cdot)$, compute the time needed to reach Γ from any starting point x. This time $\varphi(x)$ is defined by (8.5). The optimal strategy obviously consists in walking in the steepest descent direction of this time, at the prescribed speed, so that the associated desired velocity is defined by

$$U(x) = -\frac{\nabla\varphi}{|\nabla\varphi|}\, v(x). \tag{8.10}$$

8.5.   Alternative Way to Compute Desired Velocities

Some principles that are commonly used in the modeling of chemotactic phenomena can be implemented in the context of crowd motions, to compute desired velocities. In the biological context, chemotaxis refers to the ability of some living organism to move (by swimming or crawling) toward zones with higher concentration of a certain substance (called chemoattractant). When the substance is produced by the population itself, the approach leads to the Keller–Segel system (see e.g. Keller and Segel, 1971). In this context, the motion of the active population is driven by diffusion and by an extra advection term, with a chemotactic velocity along the gradient of the chemotactic agent. In the present context of the evacuation of a building, one may consider that exits emit some sort of virtual substance (like a flavor) that tends to attract people. This approach leads to solve a Laplace problem over the domain (the room, or the building to evacuate), with appropriate boundary conditions. For instance, in the case of a rectangular room with a single exit, the boundary value problem represented in Fig. 8.8 (top) leads to the desired velocity fields represented at the bottom of the same figure (bottom).

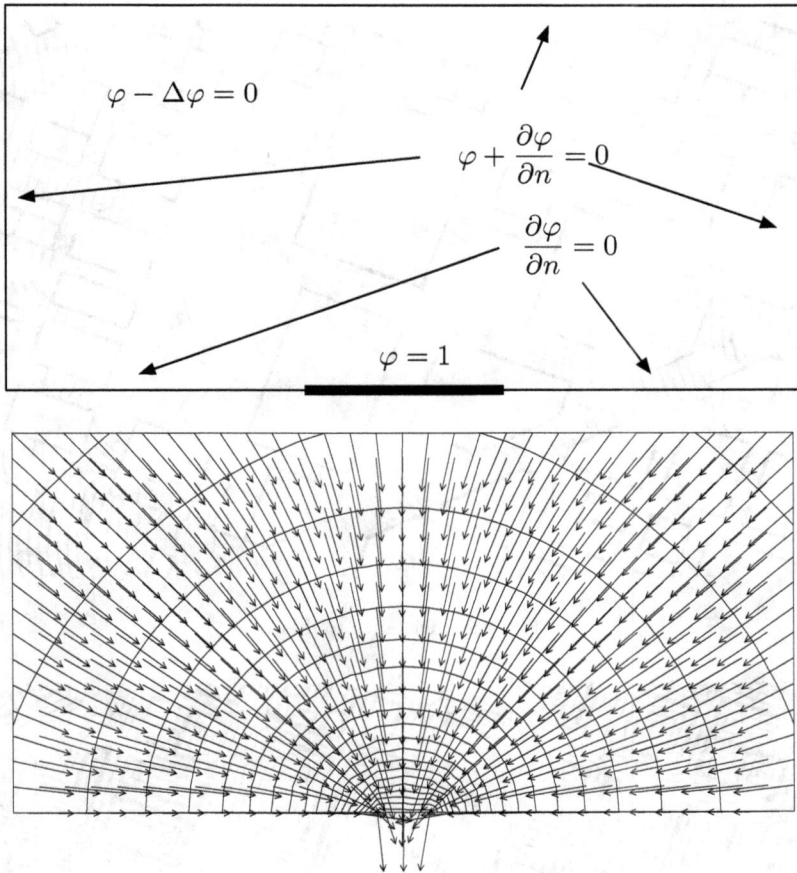

$$\varphi - \Delta\varphi = 0$$

$$\varphi + \frac{\partial\varphi}{\partial n} = 0$$

$$\frac{\partial\varphi}{\partial n} = 0$$

$$\varphi = 1$$

Fig. 8.8. Desired velocity field $\nabla\varphi/|\nabla\varphi|$.

8.6. Illustrations

The approach presented in this chapter makes it possible to rapidly compute geodesic distances together with desired velocity fields in complex geometries. Figure 8.9 represents an example of computed isovalues of the distance to some target (red segment to the left of figure), and Fig. 8.10 the isovalues of the distance to the set of obstacles.

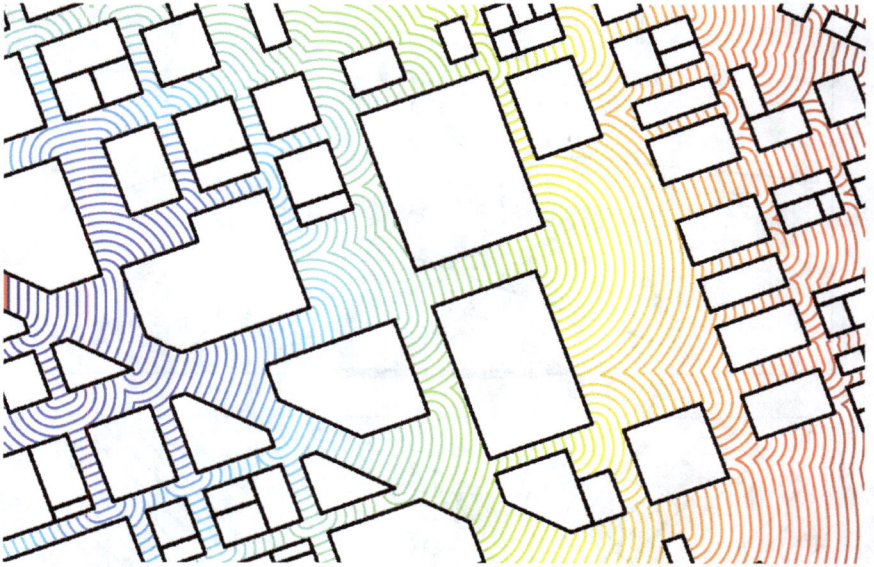

Fig. 8.9.   Distance to door.

Fig. 8.10.   Distance to obstacles.

Chapter 9

Data, Observable Phenomena

We gather in this chapter quantitative data and qualitative observations from the literature, and we describe how those characteristics are accounted for in the different models. The data presented here are of two different types, according to the way they are involved in the modeling process (upstream of downstream): *upstream* features or characteristics are used in the process of elaborating models or setting up numerical parameters, whereas *downstream* features are used to validate a modeling approach, or simply confront it to real observations. The latter class typically pertains to collective phenomena that are observed in some situations, like the Faster-is-Slower effect, or the Capacity Drop phenomena (see below). Let us insist on the fact that the meaning of "validation" differs from the one it has in more classical contexts, like particle physics, or fluid mechanics, where experiments can be set up and reproduced with a strict control of parameter values, and on the confidence that the governing rules (like Newton's law) remain the same for all experiments. As we shall see, the effects presented here do not occur systematically. As a consequence, neither reproducing them provides a quality certificate for a model, nor failing in this matter means that the model should be rejected.

9.1. Diameters

In the context of crowd modeling, individuals are commonly identified to (soft or hard) disks. Setting the diameters of those disks is a delicate issues, since most individuals seen from above present a non-circular shape.

The equivalent diameter of an individual is usually taken as the largest dimension of the real non-circular shape. It is considered to range between 50 and 70 cm (see e.g. Helbing *et al.* (2000) or Frank and Dorso (2011)). Note that by considering an homogeneous population of individuals/disks of diameter 60 cm, the individual area is $\sim 0.28\,\mathrm{m}^2$, thus a maximal density below $3.5\,\mathrm{P\,m}^{-2}$, which is significantly smaller than what is considered to be the maximal packing density for human crowds, between 5 and $6\,\mathrm{P\,m}^{-2}$ (the value $5.4\,\mathrm{P\,m}^{-2}$ is commonly found in the literature). The radius of 50 cm yields a density which is closer to the observed maximal density. Yet, if one is interested in evacuation through a bottleneck, it may be more relevant to choose a larger radius, which more properly reflects the actual width of individuals (shoulder to shoulder).

In Rouphail *et al.* (1998), a typical individual is represented by an ellipse, with minor and major axis equal to 50 and 60 cm, respectively.

In the context of cellular automata, there is a huge consensus in considering $40\,\mathrm{cm} \times 40\,\mathrm{cm}$ cells, which sets *de facto* the maximal packing density at $6.25\,\mathrm{P\,m}^{-2}$.

9.2.   Proxemics, Interpersonal Distances, Density

9.2.1.   *Proxemics and confort densities, experimental evidence*

The tendency of human beings to maintain a distance with respects to neighbors is common knowledge. The dependency of this distance upon the degree of intimacy of people, and possibly upon socio-cultural factors, has been studied in Hall (1969). Table 9.1 gathers the typical distance ranges, depending on the relationship between individuals (for US citizens).

Table 9.1.   Distances in man, from Hall (1969) (distances in cm).

Intimate Distance	–	Close Phase	$D < 15$
Intimate Distance	–	Far Phase	$15 < D < 40$
Personal Distance	–	Close Phase	$45 < D < 75$
Personal Distance	–	Far Phase	$75 < D < 125$
Social Distance	–	Close Phase	$120 < D < 210$
Social Distance	–	Far Phase	$210 < D < 360$
Public Distance	–	Close Phase	$360 < D < 750$
Public Distance	–	Far Phase	$750 < D$

9.2.2. *Proxemics, modeling aspects*

The tendency of individual to remain at a certain distance from neighbors is most commonly included in models by adding a correction term of the repulsive type (i.e. which tends to move an individual apart from his neighbors). This term has to be tuned in such a way that its effect becomes significant as soon as the actual distance becomes smaller than the distance considered to be comfortable according to the previous considerations.

For instance, in the inertial *social force model* (3.1), the effect of an individual j upon i acts along the line joining their centers, and the interpersonal distance is typically chosen as the distance δ below which the repulsive effect becomes significant.

The native form of *Cellular Automata* (Chapter 5) model does not enable a tunable account of proxemics, but one may consider that the space step (size of the cells) is considered a model variable. Adjusting this parameter Δx sets the maximal density allowed in the model, which is directly related to the minimal distance between individuals. Note also that the exclusion rule and the role it plays in the evolution step tend to produce configurations with at least a free cell between two individuals, which can be seen as a way to smoothly maintain a minimal interpersonal distance (of the order of 40 cm for the standard version of the model).

As for *granular models* (Chapter 4) the situation is similar: if one considers that the size of the rigid disk corresponds to the actual size of physical individuals, this approach does not make it possible to integrate proxemic ingredients, which are psycho-sociological in nature. Yet, if one considers that the strict constraint on inter-individual distances expresses the tendency of pedestrians to preserve a minimal distance with neighbors, setting the radius of the rigid disks at the sum of actual radius and the half of the proxemic distance amounts to prescribe the proper minimal value for distances. This setting is nevertheless questionable, because of the native symmetric character of granular interaction: in particular, it relies on the unrealistic assumption that all individuals have the same interpersonal distance, and also on the assumption that each individual instantaneously estimates their distance from all their neighbors, including from the ones *they do not see.*

9.3.   Cone of Vision

The cone of vision is defined as the zone ahead an individual, in which he is able to visually detect obstacles or other pedestrians.

9.3.1.   *Cone of vision, experimental evidence*

Some experiments have been performed (see Kitazawa *et al.*, 2008) to evaluate the maximal angle beyond which a walking individual in normal conditions does not considers objects or other pedestrians. The obtained value is around $\pi/4 = 45°$ (half angle of the cone of vision). In the conditions of the experiment, it was also estimated that the distance above which other pedestrians were considered is about 4 m.

Obviously, such a zone cannot be defined as a proper geometrical object, and it should be kept in mind that the actual zone in which an individual accounts for intrusion when walking ahead is highly dependent upon the underlying conditions (nature of the interaction with other pedestrian, state of mind, ...). In particular, if one defines the cone of vision as the zone in which the looking ahead pedestrian is *able* to detect obstacles and other pedestrians, the half angle may go up to $\pi/2 = 90°$, and the distance may reach a few hundreds of meters.

9.3.2.   *Cone of vision, modeling aspects*

The fact that a pedestrian's behavior is likely to be influenced in priority by neighbors which he sees can be included in models in a soft way or in a hard way. The soft way consists in multiplying the interaction terms by a prefactor (the term is taken from Helbing and Johansson (2009)) which depends on the location of the considered "influencer" with respect to the direction of vision, as defined for instance by Eq. (3.4). The hard way consists in adopting an all-or-nothing standpoint: the cone of vision is properly defined as a geometrical zone (as illustrated by Fig. 9.1), and a neighbor is either in or out. In the latter case, the influence they might exert is simply disregarded.

9.4.   Pedestrian Speed, Fundamental Diagram

The speed of a pedestrian, under standard conditions, is usually considered to range between 1.2 and $1.5 \, \mathrm{m \, s^{-1}}$ (see e.g Rastogi *et al.*, 2013). A huge amount of data was gathered during Work Exhibition 2000 in Hannover

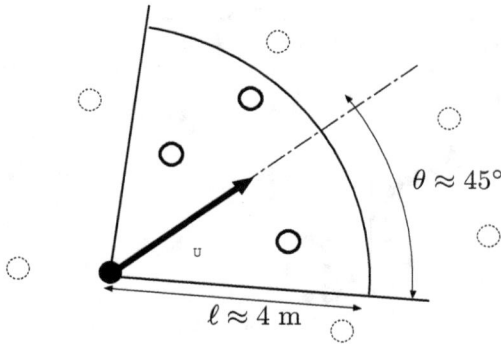

Fig. 9.1.   Cone of vision.

(see Klüpfel *et al.*, 2003). Velocities appear to be distributed in a Gaussian way about a mean value of 1.30 m s^{-1}, with standard deviation 0.21 m s^{-1}. The same authors report a significant (negative) dependence of the mean velocity upon the size of the social group he belongs to: while isolated individual had mean velocity 1.38 m s^{-1}, the velocity of 6-pedestrian groups goes down to 1.10 m s^{-1} (with a roughly affine interpolation for intermediate sizes). All those data pertain to uncongested situations.

Experiments have been performed to investigate the dependency of a pedestrian velocity with respect to the local density, in particular in the case of high densities. Figure 9.2 presents such empirical results presented in Seyfried *et al.* (2005), together with more recent experiments performed in a circular corridor at different densities (Jelić *et al.*, 2012). The average speed is represented as a function of the linear density, expressed in ms^{-1}, on the top, and as a function of the distance to the next person on the bottom. We also represent the best [distance $\mapsto$ speed] fitting curve of the type

$$w \longmapsto v(w) = U(1 - \exp(-(w - w_m)/w_s)) \quad \text{for} \ w \geq w_m.$$

The corresponding parameter values are

$$U = 1.26 \text{ m s}^{-1}, \quad w_m = 0.31 \text{ m}, \quad w_s = 0.93 \text{ m}.$$

Some authors favor a representation in terms of floor density (expressed in P m^{-2}). In this spirit, the following density-speed formula is proposed

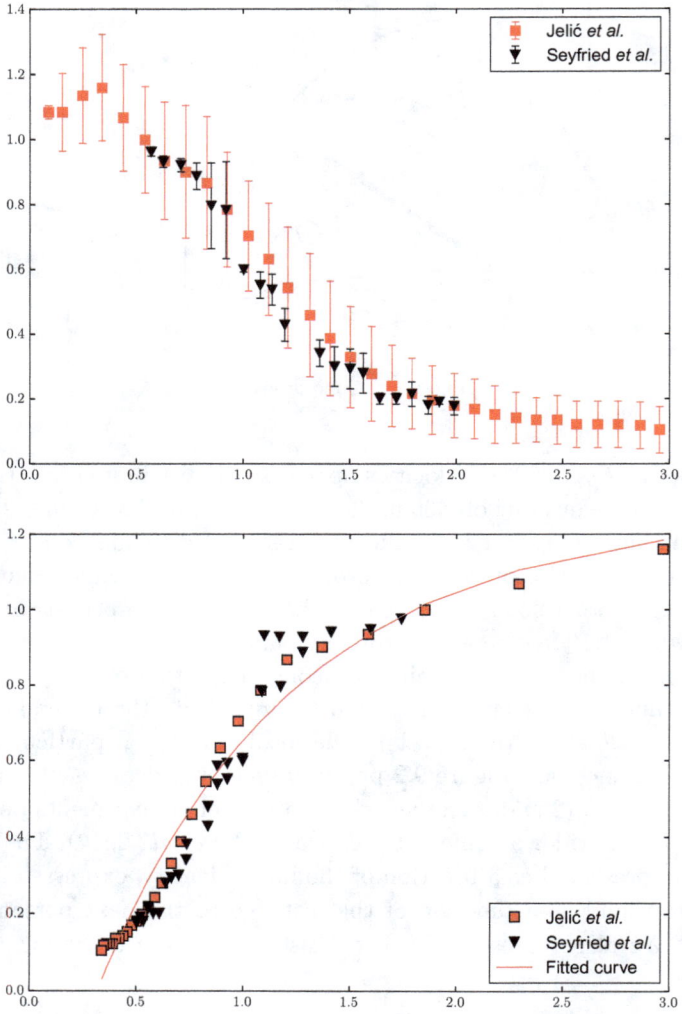

Fig. 9.2.   Speed vs. linear density (top), speed vs. distance (bottom).

in Weidmann (1992) (based on experimental data)

$$u = 1.34\left(1 - e^{-1.913\,(1/\rho - 1/5.4)}\right),$$

where ρ is the density (in $\mathrm{P\,m^{-2}}$), assumed to remain below the maximum value $\rho_{\max} = 5.4\,\mathrm{P\,m^{-2}}$.

9.5. Door Capacity

The capacity of a door is defined as the maximal number of persons who can pass it per unit time. It is expressed in $P\,s^{-1}$. The notion of capacity in the literature commonly designs the *specific* capacity, which is the maximal flux per unit width, expressed in $P\,s^{-1}\,m^{-1}$. The fact that the latter notion is favored suggests a linear dependence of the capacity upon the door width, but such a linear dependence is ruled out by many experiments. In particular, experimental data presented in Daamen *et al.* (2012) exhibit a non-monotone dependence of the (specific) capacity upon the door width. Experiments are performed for various door widths ranging over 50, 100, 160, 220, and 275 cm. The value which corresponds to the intermediate width 220 cm is much higher (around $3.5\,P\,s^{-1}\,m^{-1}$) than the values corresponding to the other width ($2.8\,P\,s^{-1}\,m^{-1}$ and lower), especially with a low stress level of the population. Typical values range between 2 and $3.5\,P\,s^{-1}\,m^{-1}$.

9.6. Capacity Drop Phenomenon

When people upstream an exit door accumulate, it has been observed in certain situations that the actual flux decreases, the actual capacity of the door drops down to a possibly much smaller value. This effect can be put in relation with the Faster-is-Slower phenomenon (see Section 9.7): when people rush to a single exit door (i.e. when they go *faster*), they tend to increase the density upstream the exit, and the capacity of the door is directly related to the speed of individual going through it, so that the slowing down of evacuees is directly related to the capacity drop. Yet those effects are usually considered as different, since the congestion near the exit zone is not necessarily due to an increase of the velocity.

9.6.1. *Capacity Drop phenomenon, experimental evidence*

Experiments are proposed in Cepolina (2009), people are asked to walk along a corridor with a fixed width of 140 cm, the width of which is reduced at some point, down to 90 or 60, depending on the experiment. The stronger effects are observed for the 90 cm restriction: when pedestrians accumulate upstream the restriction, the capacity reduces by 18 to 28%. The effect is also observable for the 60 cm restriction, but at a smaller intensity (around 15%).

9.6.2. *Capacity Drop phenomenon, modeling aspects*

Most models that handle congestion in some way reproduce the capacity drop, but this effect remains delicate to accurately quantify, and the reason why it is reproduced by a given model is usually unclear. Let us mention the extreme case of Capacity Drop recovered by the granular model (Chapter 4): in some instances, increasing the number of evacuees lead to a static jam (see e.g. Fig. 4.4, left column), so that the actual capacity of the exit door drops down to zero.

9.7.   Faster-is-Slower Effect

In the context of crowd evacuation, the so-called Faster-is-Slower effect (FiS) corresponds to a decrease of the global flow rate induced by an *increase* of people desired velocities or driving forces.

9.7.1.   *Faster-is-Slower effect, experimental evidence*

The effect has been experimentally observed in Garcimartin *et al.* (2014): the authors measured evacuation times for a population of 85 individuals under various conditions. A first set of experiments was carried out in a non-competitive spirit, and for the second set the individual were authorized to "push and jostle for the exit within reason". Exit times in the second case were systematically larger than in the first situation, with an average increase around 20%.

Another set of experiments is described and analyzed in Pastor *et al.* (2002). Experiments are performed on a herd of sheep rushing through a door, craving for food. The level of competitiveness is conditioned by the temperature, owing to the fact that the individual behavior of a sheep is strongly affected by this parameter. The time lapses between consecutive egresses of entities is observed to be significantly larger in the case with high competitiveness. The authors propose in the same paper a purely mechanical version of the experiment. The entities are now rigid grains flowing through a bottleneck, and the parameter to investigate the effect is the inclination angle of the setting. The observed effect can be described as follows: in the clogging regime, increasing the angle (and thus increasing the driving force) reduces the flow rate.

Note that this effect strongly depends on the experimental setting. Some authors do not report any significant effect (see e.g. Daamen *et al.*, 2012), where a large set of experimental data was collected to investigate the effects of various parameters upon the capacity of an exit door. In those experiments no *Faster-is-Slower* effect was observed: the increase of stress level simply leads to higher speed and higher capacity. It cannot be excluded that some other experiments may have been carried out and kept unpublished because no effect was observed.

9.7.2. *Faster-is-Slower effect, modeling aspects*

The phenomenon is reproduced by Helbing's model (see e.g. Helbing *et al.*, 2000), where the evacuation time of 200 people is computed as a function of their desired velocity. The possibility to recover this effect relies on a friction term that is added to the forcing term, which penalizes the relative velocity of people in contact. Whenever the desired velocity is increased, it leads to smaller inter-individual distances, which induces stronger frictional effects. These frictional effects tend to freeze the congested crowd, thus reducing the overall flow rate.

A large set of numerical experiments based on the same model is proposed in Escobar and Rosa (2003). The authors investigate the effect of panic in the efficiency of room evacuation. The panic is again modeled by increasing the desired velocity of individuals. It is observed that for small crowds (i.e. under 100 individuals), panic tend to speed up the process, whereas for larger crowds, increasing some desired velocities may harm it. More precisely, when the fraction of "panicked" people exceed a certain value, a decrease of the evacuation efficiency can be observed.

In a fully different setting, the granular approach (see Maury and Venel (2011), or Chapter 4 of the present book) is also able to recover this effect, whereas no friction is accounted for. Let us first state that a straight increase of the desired velocities modulus is ruled out as a way to recover this effect (see Section 10.1). An alternative protocol is thus proposed in Faure and Maury (2015): individuals have the ability to reduce their own desired velocity modulus whenever they are in contact with some neighbor in front of them, so that pushing forward is inefficient. Adding this simple behavior to the model leads to a drastic fluidification of the overall flow. Those considerations are detailed from a mathematical standpoint in Section 10.1.

Cellular automata models (presented in Chapter 5) are known to reproduce the Faster-is-Slower effect under some conditions. The crucial ingredient is *friction*. In this context, as detailed in the dedicated chapter, friction is implemented by modifying the handling of conflicts, in the parallel update algorithm. All desired motions of particles are drawn simultaneously, which unavoidably leads to conflict: two or more particles target the same cell. Adding friction amounts to consider that there is a probability $\mu > 0$ that the conflict is resolved by leaving all competitors in their cell: nobody moves. Here lies the *Slower* feature. The *Faster* ingredient is implemented by considering that the parameter κ which conditions the transition probabilities (see Eq. (5.2)) quantifies the eagerness of individual to move forward. In this context, it can be checked that, if μ is positive, increasing κ (i.e. having agents try to move faster) may lead to a decrease of the evacuation efficiency. The mechanism here is quite clear: increasing κ tends to increase the number of conflicts, some of which (proportion μ) are fully counter-productive. A toy problem is proposed in Section 10.1 to formalize this effect of friction in the context of cellular automata.

Large sets of numerical experiments based on CA are presented in Xiaoping *et al.* (2010). The total evacuation time for 200 virtual individuals is computed for several desired speeds. This evacuation time presents a minimum for a speed around $1.5\,\mathrm{m\,s}^{-1}$, and then increases to higher values (about 20% higher for a speed of $4\,\mathrm{m\,s}^{-1}$).

Remark 9.1. The FiS effect is sometimes put in relation with the so-called Yerkes and Dodson law (see e.g. Wang, 2016) which essentially states that, while moderate stress improves human performance, excessive stress may impair it. We must stress here that the FiS effect which we previously described is *collective* in essence. Indeed, the microscopic models we described are *not* based on any Yerkes–Dodson ingredient at the individual level. Under (possibly excessive) stress, individuals are assumed to develop higher capacities (force, velocity). Only their interactions may lead to a counter-productive effect of excessive forcing. To conclude this remark, let us mention that Yerkes–Dodson ingredients might be implemented in microscopic models, e.g. by considering that, beyond a certain level of stress, individuals start to erratically behave, and become incapable of developing reasonable (at both individual and collective levels) strategies.

9.8.   Influence of an Obstacle

It is commonly heard that placing an obstacle upstream an exit door contributes to significantly fluidize the pedestrian motion, and thus decrease the evacuation time during emergency exit. This paradoxical effect is delicate to reproduce and observe, since it is expected to occur in real panic situations, when people tend to push against each other. As a consequence experimental evidence of this effect for human crowds is rare. "Alternative evidence" is proposed in the literature, based on experiments implying non-human living entities (like sheeps, ants, mice), or based on in-silico experiments (mainly microscopic models). At any rate, this effect should not be considered as systematic and robust; its occurrence highly depends on the configuration (size of the exit, shape and position of the obstacle), and also on the behavior of agents (politeness/competitiveness, tendency to develop individual or collective strategies).

9.8.1. *Influence of an obstacle, experimental evidence*

An experimental setting is described in Yanagisawa *et al.* (2009), based on pedestrians going through a 0.5 m exit. Two series of experiments are described, a first one without obstacle, and a second one with a cylindrical obstacle (case (a) in Fig. 9.3) located asymmetrically with respect to the exit. Experiments exhibit a 4% increase of the flux by addition of the obstacle. This small increase is reported as statistically significant by the authors.

In Jiang *et al.* (2014), the effect of pillars placed upstream a 1 m wide exit is investigated, numerically (social force model, see next paragraph) and experimentally. From the experimental standpoint, the mean flux is about 16% larger for the case with two obstacles ($2.9 \, \mathrm{P \, s^{-1}}$), compared to the situation with no obstacle ($2.5 \, \mathrm{P \, s^{-1}}$). The number (three) of experimental runs for each case does not make it possible to give confidence intervals.

In Helbing *et al.* (2005), various experiments are presented. In particular, the authors consider the evacuation of a room in panic-like situation. Placing a board of width 45 cm in front of the exit (of width 82 cm) (the distance to the exit is not given) is shown to reduce the evacuation time by 30%.

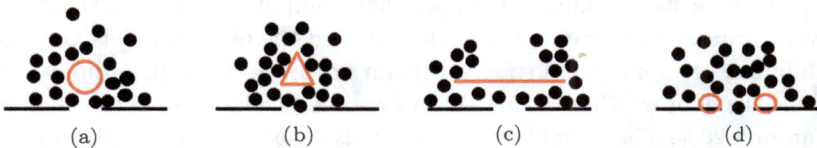

(a)   (b)   (c)   (d)

Fig. 9.3.   Obstacles.

More significant effects have be established for non-human entities. In Lin *et al.* (2017), experiments based on mice are described. Mice (crowds of 80 entities were considered) are driven to pass through an exit, with or without obstacle. The presence of an obstacle is shown to reduce the evacuation time by a maximum of 36% (the effect varies with the position of the obstacle). The maximal effect is obtained for an obstacle placed at 4 cm from the exit, while mice are typically 3 cm wide and 10 cm long.

Experiments proposed in Zuriguel *et al.* (2016) involve sheeps. The authors investigate the effect of placing a cylindrical obstacle (with diameter 114 cm) upstream a 96 cm exit (distance to the exit ranging from 60 cm to 1 m), case (a) in Fig. 9.3. The width of a sheep is about 35 cm. They report a positive effect of the obstacles for distances of 80 and 100 cm (for 60, the obstacle is counter-productive).

9.8.2. *Influence of an obstacle, modeling aspects*

The fluidizing character of an obstacle can be recovered by models of the granular type, like those presented in Chapter 4, see Fig. 4.4. The presented numerical tests correspond to an extreme situation: a static jam is created during the standard evacuation, whereas adding an obstacle leads to a full evacuation.

The social force model is also capable of recovering this effect, as presented in Frank and Dorso (2011). In this paper, the authors investigate the effect of an obstacle upstream the exit within this framework. The door width is set at $L = 1.2$ m, and diameters of individuals are uniformly distributed between 0.5 and 0.7 m. Two types of obstacles are considered: a cylindrical pillar of diameter L, and a thin flat panel of length $4L$ (cases (a) and (c) in Fig. 9.3).

In Starke *et al.* (2014), the authors investigate the effect of a triangle-shaped obstacle on the flow rate through a exit door (case (b) in Fig. 9.3), computed with a model of the social force type. They investigate the dependence of the flux upon obstacle parameters, like position and shape. In particular, for a fixed triangular shape, they compute the flux associated to several values of the gap between the exit and the obstacle. This approach exhibits a clear maximum for a certain value of the gap, around 1.2 m (while the door width is 0.9 m). At this regime, the increase in terms of flux is around 20%. The social force model is also used in Jiang *et al.* (2014) to investigate the effect upon evacuation of two pillars on the sides of the main outflow stream (case (d) in Fig. 9.3). It is observed that such obstacles can fluidize the flow by reducing the *tangential momentum*, which is defined as the sum of the interaction forces (in absolute value) along the wall direction. This tangential momentum quantifies in some way the *stress* (in a mechanical sense) in the transverse direction; it vanishes for an unperturbed stream of pedestrian walking in the same direction, and tends to increase when a bottleneck is met. The optimization is performed using a genetic algorithm. For two obstacles, the increase of the outlet flux goes up to 40%.

Cellular automata models can also reproduce, under some circumstances (the position of the obstacle is in particular highly sensitive) an increase of the mean outflow by addition of a small obstacle upstream the exit door. Numerical evidence is presented in Yanagisawa *et al.* (2009), with a CA model of the parallel type, with friction (we refer to p. 88 for the meaning of friction in the context of cellular automata). The authors observe that the standard accounting of friction does not reproduce the positive effect of the obstacle. Yet, by addition of two ingredients, namely (1) a *frictional function*, which makes it possible to account for the number of individuals involved in a conflict, and (2) a *turning function*, which encodes inertial effects from an angular standpoint (difficulty for a pedestrian to suddenly change their direction), the authors recover an increase in the evacuation flux similar to what is observed in their experiment (around 4%).

In Xiaoping *et al.* (2010), the authors investigate the effect of a partition wall upstream the exit on evacuation times (case (c) in Fig. 9.3). CA computations indicate that, for a small number of evacuees, increasing the width of this wall tends to reduce the evacuation time, whereas for larger numbers of individuals, increasing the width first increases the evacuation time, then reduces it. The computed reduction of evacuation times ranges typically between 10% and 20%.

9.9. Stop-and-Go Waves

In the context of crowd motions, the term "Stop-and-Go" refers to an alternation of slowdown (possibly down to a complete stop) and acceleration phases experienced by a pedestrian embedded in a large crowd of individuals heading to the same direction. It therefore pertains to a Lagrangian description of the motion, whereas the term "waves" pertains to an Eulerian description of the crowd: it describes the upstream (opposite to the movement direction) propagation of Stop-and-Go zones.

9.9.1. *Stop-and-Go waves, experimental evidence*

This phenomenon has been observed and quantified experimentally in various instances. In Portz and Seyfried (2011), the authors describe experiments performed in a 26 m long circular corridor, with various numbers of pedestrians (up to 70). Stop-and-Go waves were observed (spontaneous emergence) as soon as the number of pedestrians exceeded 45.

A similar experiment is proposed in Lemercier *et al.* (2016), with path lengths between 15 m (inner circle path) and 26 m (outer circle path), population number taking various values between 8 and 28.

9.9.2. *Stop-and-Go waves, modeling aspects*

In Portz and Seyfried (2010), the authors propose an event-driven modification of Helbing's model which reproduces SaG waves. The approach is based on the notion of *safety distance* that depends on the current velocity. A pedestrian accelerates to their desired velocity until the distance from the individual ahead is smaller than this safety distance, and then decelerates until the distance is larger, and so on. In this approach, the time step plays the role of a reaction time, therefore it cannot be seen as a natively differential model, since the actual behavior depends on how this parameter is chosen.

SaG waves are also reproduced by a model proposed in Lemercier *et al.* (2016), that is inspired by a model initially proposed in Aw *et al.* (2002) for road traffic. In this setting, the acceleration of a pedestrian is written as a function of the relative velocity with respect to the pedestrian in front, at some previous instant, with a factor which depends on the local density.

9.10. Further Considerations on Human Behavior

The case of emergency evacuations has been taken as a generic example in this book, for the simple reason that in such a situation people's tendency may seem easy to determine: instantaneously trying to reach the closest exits as soon as possible sounds like a reasonable goal. Yet, it has been reported that, in real emergency situations, people may react differently. For instance, it has been reported that the time at which someone is made aware of a real danger is usually followed by a certain prostration time during which the person is actually incapable of deciding anything, significantly delaying the start of the evacuation procedure. Furthermore, the natural tendency of a person trying to evacuate a complex building is not systematically to approach the closest exit (which is usually pointed to by pieces of signage), but rather to walk their incoming way (the one they know) backward (see Levin, 1984).

Chapter 10

A Wider Look on Characteristic Phenomena in Crowds

This chapter proposes abstract developments aiming at identifying which ingredients in a model are likely to lead to some particular phenomena which are typical of crowd motions, namely the Faster-is-Slower(FiS) effect, the fluidizing role of an obstacle, and Stop-and-Go waves.

10.1.   Faster-is-Slower Effect

The FiS effect belongs to a larger class of phenomena, all of which are characterized by the following feature: a forcing term triggers a response of a certain system, in a faithful way in the sense that only the null action induces a null response, but monotonicity is lost, at least beyond a certain intensity of the forcing term. In other words, we are interested in systems which illustrate the popular saying:

What is better is the enemy of what is good.

The FiS effect assessed by some experiments (see Section 9.7) clearly fits in this class, but the very reason why it does so is not clearly identified. This uncertainty motivates the developments proposed in this section: which characteristics of a forced system might lead to such a paradoxical behavior?

Let us start by describing systems (some of which are used in the context of crowd modeling) which *may not* exhibit such a behavior. We will present

161

them in the context of crowd motion, but they may be relevant in many other contexts. Let us denote by $U \in \mathbb{R}$ a parameter which quantifies the desired velocities of people in an evolution model of the type

$$\frac{dq}{dt} = F_U(q(t)),$$

where $q(t) = (q_1(t), \ldots, q_N(t))$ denotes the positions of individuals at time t. Assume that the mapping $U \longmapsto F_U(q)$ is linear, for any configuration q. Of course, the mapping $q \longmapsto F_U(q)$ may be highly nonlinear, possibly non-smooth (like in Chapter 4), and non-local in the sense that $F_U(q)_i$ (that is the actual velocity of individual i) may depend on positions of all other individuals. Then increasing the desired velocities amounts to multiply U by a factor $\lambda > 1$. It follows that, if $t \mapsto q(t)$ is a solution to the initial problem, then $q_\lambda(t) = q(\lambda t)$ is a solution to the problem associated to velocity λU. As a consequence, increasing the desired velocity simply speeds up the evolution process by a factor λ, and the evacuation time is accordingly *reduced* by a factor $1/\lambda$ (i.e. no FiS effect).

As a consequence, recovering this effect calls for including some nonlinearity with respect to the parameter which quantifies the desired velocity in the model, or some other ingredients like the ones introduced below.

Mechanical friction

We consider the following (non-inertial) mechanical system: an object (the disk in Fig. 10.1) is subject to a diagonal force (λ, λ). It pushes upward and rightward on a horizontal bar with resting position $y = 0$. The bar is assumed to exert a pullback (downward) force which is a function φ of y. On the other direction, a dissipative force opposes the horizontal motion

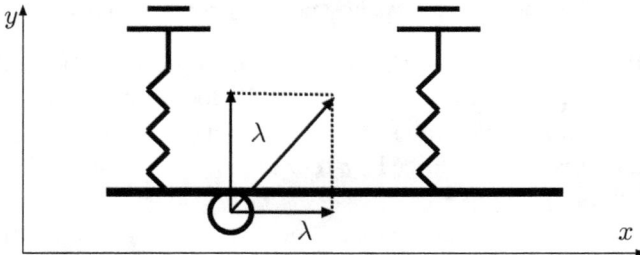

Fig. 10.1.   Toy mechanical model.

with a coefficient that is a function of y: the more the disk pushes on the bar, the larger the friction coefficient is. We seek a stationary state for this system, which corresponds to the disk moving at a constant height y, at a constant horizontal speed u. The vertical force balance reads $y = \varphi(\lambda)$, whereas, horizontally, the dissipative viscous force $-\mu(y)u$ balances λ, i.e. $\lambda - \mu(y)u = 0$. It yields

$$u = \frac{\lambda}{\mu(\varphi(\lambda))}.$$

In the previous equation, λ quantifies the action exerted on the disk, which tends to move it rightward, and u is the response of the disk (horizontal velocity). Now, introducing the response function $\psi = \mu \circ \varphi$,

$$\frac{du}{d\lambda} = \frac{\psi(\lambda) - \lambda\psi'(\lambda)}{\psi(\lambda)^2},$$

which is negative (FiS effect) as soon as $\psi(\lambda)/\lambda < \psi'(\lambda)$. Figure 10.2 illustrates two typical situations:

(1) The function ψ vanishes at 0, with a non-negative slope, and it is convex (see Fig. 10.2, left). In that situation the FiS effect is extreme. The slope of the chord is smaller to the slope of ψ, and the maximal velocity (which is $1/\psi'(0)$) is approached by having λ go to 0.
(2) The function ψ is strictly positive at 0, and it is strictly convex. In this situation, the actual velocity increases with λ for small values (see Fig. 10.2, right), and then (for $\lambda > \lambda_c$) it becomes counterproductive to further increase the forcing.

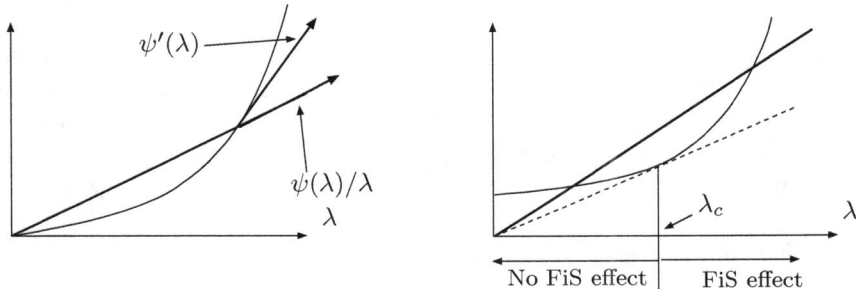

Fig. 10.2. Velocity as a function of the distance.

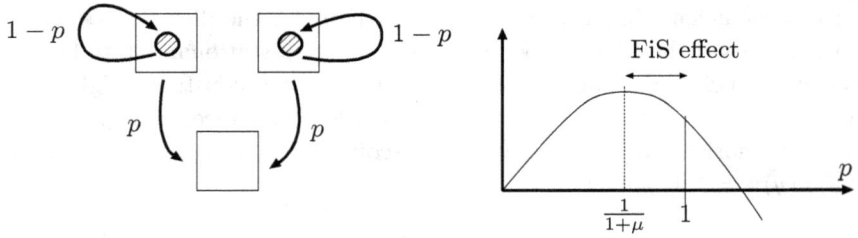

Fig. 10.3.   FiS effect for a toy cellular automaton with friction.

Stochastic friction (cellular automata setting)

The FiS can be recovered by the cellular automata model, as soon as friction is implemented. Let us recall that, in this context (see Chapter 5), adding friction consists in considering that there is a probability $\mu > 0$ that, in the event of a conflict (two or more particles compete for the same cell), no one moves. Let us consider the simplest situation one may imagine to understand how friction in this sense is likely to reproduce FiS effect. Consider two particles, each in their own cell, competing for the same empty cell.

More precisely, each particle has a probability $p > 0$ to attempt a jump in the third cell, and both attempts are drawn independently (see Fig. 10.3, left). If only one of them attempts a jump (probability $2(1 - p)p$), it is realized, and we affect the score 1 to this outcome. If none attempts a move, (probability $(1 - p)^2$), no one moves and the outcome is 0. If both try a jump (probability p^2), there is a probability μ that no one moves (outcome 0), and the complementary probability $(1 - \mu)$ that one of them is authorized to move. To sum up, the expected value of the outcome is

$$\alpha = 2(1 - p)p \times 1 + 0 \times \mu(1 - p)^2 + \mu p^2 \times 0 + (1 - \mu)p^2 \times 1$$
$$= 2(1 - p)p + (1 - \mu)p^2 = p\left(2 - (1 + \mu)p\right),$$

which is maximal for $p_{\max} = 1/(1 + \mu)$ (see Fig. 10.3, right). As a consequence, as soon as $\mu > 0$, the maximal outcome (maximal mean speed) is obtained for this $p_{\max} \in (0, 1)$. In other words: beyond this value, increasing p (i.e. increasing the desired speed of agents) tends to decrease the (expected value of the) effective motion.

Steepest descent in a non-convex setting

We consider here a very simple system of the gradient flow type, with control parameters, to illustrate to the fact that, when convexity of interaction potentials is ruled out, it may be efficient to *slow down* in some way to approach *more rapidly* one's objective. In order to preserve simplicity, we sacrifice the direct connection with crowd motions, and we refer to Faure and Maury (2015) for similar developments in direct connection with evacuation scenarios. The setting we consider is the following: the "crowd" is made of two individuals at x_1, $x_2 \in \mathbb{R}$, initially located at the same location $x^0 > 0$, each of which tends to decrease their dissatisfaction $D(x) = x$. The associated gradient flow would consists of two identical (and independent) leftward motions at constant velocity -1. Now consider an interaction (non-convex) potential defined as

$$V(x_1, x_2) = \kappa e^{-(|x_1|^2 + |x_2|^2)/\delta^2},$$

and the gradient flow associated to this potential, partially controlled in the sense that each individual has the ability to lower their spontaneous velocity. We thus define $\beta(t) = (\beta_1(t), \beta_2(t)) \in [0, 1]^2$, the collection of strategies (or controls), and we consider the generalized gradient flow (generalized in the sense that the potential depends on the control variable β)

$$\frac{dx}{dt} = -\nabla(\beta_1 x_1 + \beta_2 x_2 + \kappa e^{-(|x_1|^2 + |x_2|^2)/\delta^2}).$$

In the previous expression, the second term accounts for strong interaction, on which agents have no control, and $|x_i|$ corresponds to the quantity that each i tends to decrease. The parameter β_i, when smaller than 1, expresses the ability of agents to take on themselves and decrease the importance of their own dissatisfaction. Having $\beta_i < 1$ implements a reduction of i's spontaneous velocity. Now consider the straight gradient flow, with $\beta \equiv (1, 1)$, starting from (x^0, x^0), with x_0 significantly larger than δ. The point $(x_1(t), x_2(t))$ follows a straight path toward $(0, 0)$, and converges to a critical point where spontaneous tendencies balance the interaction term. The global dissatisfaction decreases down to a limit value which is positive (both agents remain fairly unsatisfied). If one considers now the situation where 1 momentarily reduces its weight, by having β_1 decrease to 0 or a small value, the couple circumvents the smooth obstacle, which leads to a much more favorable evolution (both dissatisfactions become negative). The two evolution paths are represented on Fig. 10.4. Note that reducing β_1 amounts to reduce the speed of the couple, which is $\sqrt{\beta_1^2 + \beta_2^2}$, and this reduction of the speed eventually leads to a more efficient (and faster) reduction of the dissatisfaction. This highlights in an abstract way the possibility to recover a *slower is faster* in a very simple modified gradient flow

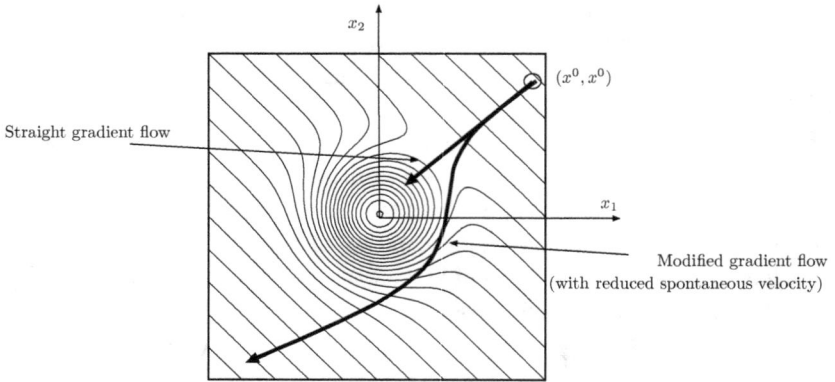

Fig. 10.4.   Gradient flow setting.

framework. We emphasize the importance of the non-convex character of the interaction potential $V(x_1, x_2)$.

Back to crowd motions

The toy problem presented above shares some features with crowd motion models for evacuation, in particular models of the gradient flow type like those presented in Section 3.2. In this context, the evolution is determined as a gradient flow for a global dissatisfaction functional:

$$\frac{dx}{dt} = -\nabla \Psi(x) = -\nabla \left(\sum_i V_i(x_i) + \sum_{i \neq j} V_{ij}(D_{ij}) \right), \qquad (10.1)$$

where the individual potentials V_i are typically defined as the distance to the exit, i.e. $V_i(x_i) = D(x_i)$. This function is typically convex (e.g. in the case of a convex room with a single exit). As a consequence, the gradient flow based on the sole potential $\sum_i V_i(x_i)$ does not exhibit any FiS effect. More precisely, if one includes the possibility for agents to diminish their velocity by considering the controlled gradient flow

$$\frac{dx_\beta}{dt} = -\nabla \left(\sum_i \beta_i V_i(x_i) \right), \quad \beta_i \in [0, 1],$$

reducing the coefficients β_i (i.e. some agents go *slower*) is obviously counterproductive, in the sense that the value of the associated dissatisfaction is always larger than the one obtained for the straight gradient flow. Now, accounting for interactions amounts to add the second term in the right-hand side of (10.1). Under quite general assumptions, e.g. if one considers that $V_{ij}(w)$ is a decreasing function of w, this potential is *not convex* with respect to $x = (x_1, \ldots, x_N)$, which induces a possible *Faster-is-Slower* effect (or, more precisely, its reversed equivalent *Slower-is-Faster* effect).

Positive vs. negative feedback in the respiratory process

We borrow here a model from a domain which may seem quite remote from crowd motions, namely the respiratory process, and more precisely the so-called Flow Limitation. As presented in Tantucci *et al.* (2002), the expiratory flow is subject to remain below a threshold value, which is *effort-independent*. It means that, above a certain value, increasing the expiratory effort is counterproductive, or at least does not increase the peak expiratory flow. Some elements have been given in the literature to explain this phenomenon, based in particular on Fluid Dynamic considerations (see e.g. Hyatt *et al.*, 1981, where the Bernoulli's law is used to model this phenomenon). We shall not address here this delicate modeling issue, but rather consider a toy model which contains minimal ingredients to reproduce this sort of phenomenon. Consider a balloon connected to a tube (see Fig. 10.5, left), which corresponds to the respiratory tract. The free end of the pipe is in contact with the outside world, set at pressure 0 (mouth and nose). The whole system is immersed in a medium (the thoracic cage) set at some positive pressure P. We assume that the pressure on the balloon is P, and that the flow Q through the pipe is given by the Poiseuille's law $P - 0 = RQ$. The resistance R is a decreasing function of the diameter of the pipe (large resistance for small diameters). We assume that the pipe is deformable, i.e. that its diameter is an increasing function of the pressure

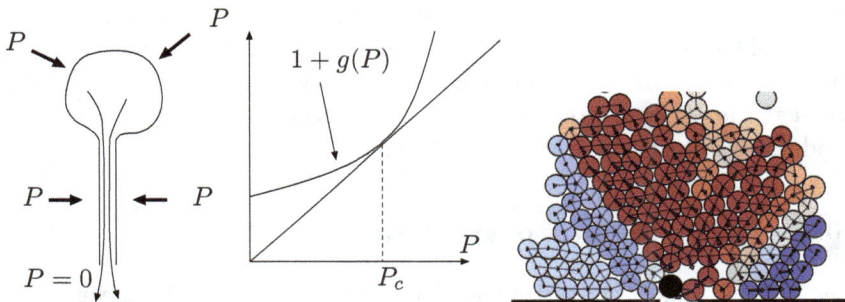

Fig. 10.5.   Maximal expiratory flow vs. crowd evacuation.

drop between the inside and the outside. We shall actually consider that it is proportional to the average value of the inside pressure, which we estimate at $P/2$, and the outside value, that is P. To sum up, we write the resistance as

$$R(P) = R_0(1 + g(P)),$$

where R_0 is the resistance associated to the pipe at rest $(P = 0)$, and $p \mapsto g(P)$ is an increasing smooth function, with $g(0) = 0$. When P increases, the pressure drop tends to push the fluid through the pipe, but it also compresses the pipe, thereby *increasing its resistance*, which tends to reduce the flow. More precisely

$$Q = \frac{P}{R} = \frac{P}{R_0(1 + g(P))} \implies \frac{dQ}{dP} = \frac{1}{R_0}\frac{1 + g(P) - Pg'(P)}{(1 + g(P))^2}.$$

The derivative is positive for small values of P, which means that a small effort will induce a positive expiratory flow. But if $P \mapsto 1 + \varphi(P)$ is strictly convex, the derivative becomes negative when P is larger than a critical value P_c, with

$$\frac{1 + g(P_c)}{P_c} = (1 + g(P_c))',$$

which corresponds to the point at which the derivative is equal to the chord (see Fig. 10.5, middle). For pressure values larger that P_c, the effect of further increasing the effort is counterproductive, because the downward effort is automatically accompanied by a lateral compressing effort which increases the resistance of the pipe. This effect can be put in parallel with the investigation of the FiS effect for the granular model (see Chapter 4). We reproduce in Fig. 10.5 (right) the snapshot of an evacuation. The red color right upstream the door correspond to the standard positive effect: the efforts of the agents behind tend to push downward the people in front, whereas the effort exerted by people on the sides (blue color) is counterproductive in terms of evacuation efficiency.

10.2. Fluidizing Effect of an Obstacle

The fluidizing effect of an obstacle placed upstream an exit has been widely commented in the literature, but it is delicate to reproduce in actual experiments and simulations (see Section 9.8). Like for the FiS effect presented in the previous section, there is no consensus on the possible causes of such an effect, and we shall restrict ourselves to general remarks on features which may induce such a paradoxical behavior. Let us start by emphasizing the paradoxical character of this phenomenon, by a very simple remark. Consider a functional $J : X \longrightarrow \mathbb{R}$ defined on a set X, and assume that J admits an infimum over X (possibly equal to $-\infty$). Assume now that the infimum is restricted to those elements x that verify some constraints (those constraints play the role of an obstacle), which reduces the feasible set to some subset $X' \subset X$. By definition of the infimum, it obviously holds that

$$\inf_{x \in X'} J(x) \geq \inf_{x \in X} J(x),$$

which expresses the obvious fact that constraints (i.e. obstacles) hinder the minimization procedure.

Global optimization vs. local optimization

Reproducing the paradoxical effect calls for an enrichment of the setting. For this reason, we again consider a dynamic situation of the gradient flow type: $t \mapsto x(t)$ is a moving point in $\mathbb{R}^2$, which evolves along the steepest descent direction of a convex functional Ψ:

$$\frac{dx}{dt} = -\nabla \Psi(x).$$

Isovalues of the functional are represented in Fig. 10.6 (left), together with the unconstrained gradient flow starting from some initial point. The corresponding path converges to the minimum of Ψ over $\mathbb{R}^2$. Assume now that the point x is subject to remain outside some region (union of the two dashed ellipse-shaped bodies represented in Fig. 10.6, middle). The

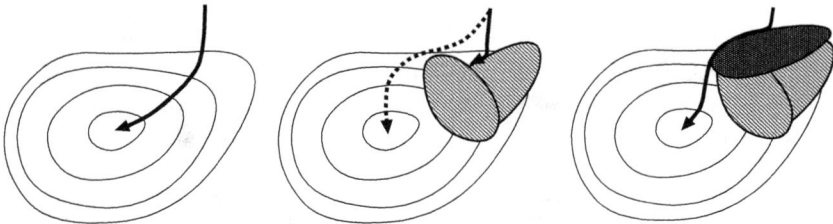

Fig. 10.6.   Constrained gradient flow.

constrained gradient flow can be written[1] as

$$\frac{dx}{dt} = P_C(-\nabla\Psi(x)),$$

where P_C is the projection on the cone of feasible directions (i.e. directions pointing inward the admissible domain). The corresponding path is *trapped* within a local minimum of Ψ in the feasible domain. Informally, one may say that heading to the local steepest descent direction is a very bad strategy to actually minimize the value of the functional, in the long run.

A smaller (i.e. better) value can be attained by following the path represented as a dotted line in the same figure. This would correspond to a strategical behavior, which calls for a full knowledge of the overall landscape. Yet, the situation can be improved without sacrificing the gradient flow character of the evolution process, i.e. keeping the evolution principle based on local optimization of the direction, by adding an *extra obstacle*, like the one represented in Fig. 10.6, right. The corresponding path is locally worse than the unconstrained one (during the time the point flows along the extra obstacle), but the final outcome is better: the path converges to the global minimum of Ψ. Note that the possibility to improve the long-term behavior by adding an obstacle is due to the *non-convex* character of the feasible set. In this spirit, we refer to Fig. 4.4 for an example of an evacuation scenario. In the left column, the evacuation (without obstacle) leads to a static jam, whereas with an obstacle (right column) the evacuation proceeds until everyone is out. Let us stress that the considerations presented in this section do not claim to prove in any way that adding an obstacle systematically improves the evacuation, neither that the non-convexity of the feasible set is the only reason for which such a phenomenon can be observed.

10.3. Damping, Propagation, and Stop-and-Go Waves

We gather here some general remarks concerning the behavior of solution to Ordinary Differential Equations around equilibrium points. We aim at identifying general features which may lead to damping, propagation, and possibly spontaneous emergence of instabilities in multi-agent systems.

---

[1]A more rigorous presentation of this constrained gradient flow calls for the notion of *Fréchet subdifferential*, and we refer the reader to Chapter 4 for a detailed description of this setting.

First order in time models

Let us start with a general autonomous system in $\mathbb{R}^N$, of the type

$$\frac{dw}{dt} = F(w).$$

We suppose that this system admits an equilibrium point W_{eq}, i.e. such that $F(W_{\text{eq}}) = 0$. By Theorem A.2, this point is asymptotically stable as soon as all eigenvalues of $\mathbf{M} = \nabla F(W_{\text{eq}})$ have a negative real part. The archetypal (and simplest) example is the ODE $\dot{x} = -\mu x$. In Chapter 2, we presented a more complex system of the type

$$\frac{dx_i}{dt} = \varphi(x_{i+1} - x_i),$$

with (in the periodic setting) $N + 1 \equiv 1$. Expressed in terms of distances (considering that the positions evolve in a periodic ring of length L), the system takes the general form

$$\frac{dw_i}{dt} = \varphi(w_{i+1}) - \varphi(w_i).$$

It obviously admits an equilibrium point $W_{\text{eq}} = (w_{\text{eq}}, \ldots, w_{\text{eq}})$, with $w_{\text{eq}} = L/N$. We write the gradient of F at w_{eq}:

$$\nabla F(w_{\text{eq}}) = \varphi'(w_{\text{eq}}) \begin{pmatrix} -1 & 1 & 0 & \cdot & 0 \\ 0 & -1 & 1 & \cdot & \cdot \\ \cdot & \cdot & \cdot & \cdot & 0 \\ \cdot & \cdot & \cdot & -1 & 1 \\ 1 & \cdot & \cdot & 0 & -1 \end{pmatrix} = \varphi'(w_{\text{eq}})(-\mathbf{I} + \mathbf{C}),$$

where $\mathbf{C}$ is a circulant matrix, with eigenvalues located on the circle of $\mathbb{C}$ centered at -1, with unit radius. We consider the situation where the model is relevant in terms of behavior: φ is assumed to be increasing, which implements the fact that, when a distance $x_{i+1} - x_i$ decreases, the velocity of i decreases as well, which tends to counterbalance the decrease in distance. As detailed in the proof of Proposition 2.6, the eigenvalue 0 is irrelevant here, because the system is overdetermined. As a consequence, all relevant eigenvalues have a negative real part, which characterizes asymptotic stability, more precisely exponential dumping with various characteristic times.

Beyond this structural stability (it only depends on the monotonicity of φ), the form of the gradient explains the tendency of the system to propagate perturbations in the upstream direction (see Section 2.1). Note that this upstream propagation can also be explained in the non-periodic setting, with completely different arguments (the matrix is no longer diagonalizable). The combination of those two features (stability and propagation) makes this kind of systems able to reproduce damped Stop-and-Go waves. Indeed when a perturbation propagates along the line of pedestrians, each of those undergoes periods of small distances (small velocities) and large distances (large velocities). But this setting does not make it possible to recover the spontaneous emergence of instabilities (such an emergence would be in contradiction with asymptotic stability). Such effects call for new ingredients.

Second order in time models and instabilities

Let us start with the simple example $\dot{x} = -\mu x$, where $x(t) \in \mathbb{R}$. A second order in time model can be elaborated by simply writing

$$\ddot{x} + \frac{1}{\tau}(\dot{x} + \mu x) = 0,$$

which is the damped harmonic oscillator equation. The equation can be written as a first-order system in the variables (x, v) (with $v = \dot{x}$), it admits a single equilibrium point. The eigenvalues of the linearized problems are

$$\lambda^{\pm} = \frac{1}{2\tau}(-1 \pm \sqrt{1 - 4\mu\tau}).$$

The system may present an oscillatory behavior (as soon as $4\mu\tau > 1$), but those oscillations are always damped. We shall now proceed to more complex system (like the inertial FTL model), obtained from stable first-order in time models, and explain why such systems may exhibit instable behavior under some conditions, as soon as some eigenvalues of the first-order system have *large imaginary parts* compared to their real parts (in absolute value). In this spirit, we consider again a first order in time system $\dot{w} = F(w)$ which admits an equilibrium point $W_{\text{eq}} \in \mathbb{R}^N$. We furthermore assume that, like for the FTL model in the periodic setting, all eigenvalues of the gradient of F at W_{eq} have negative real parts (asymptotic stability). We now consider the "inertial" version of this system:

$$\ddot{w} + \frac{1}{\tau}(\dot{w} - F(w)) = 0,$$

which can be written on the variables (w, v), with $v = w$. It admits the equilibrium point $(W_{\text{eq}}, 0)$. The stability of this point depends on the properties of the matrix

$$\begin{pmatrix} 0 & \mathbf{I} \\ \dfrac{1}{\tau}\mathbf{A} & -\dfrac{1}{\tau}\mathbf{I} \end{pmatrix}, \tag{10.2}$$

with $\mathbf{A} = \nabla F(W_{\text{eq}})$. The eigenvalue problem consists in finding λ and $(w, v) \neq (0, 0)$ such that

$$v = \lambda w \quad \text{and} \quad \mathbf{A}w - \frac{1}{\tau}w = \lambda v.$$

We denote by (μ_k, w_k) the eigenelements of $\mathbf{A}$. For each k, the new problem admits two eigenvalues $\lambda_k^{\pm}$, with associated eigenvectors $(w_k, \lambda_k^{\pm}w_k)$, where the $\lambda^{\pm}$ are the solutions to

$$\lambda^2 + \frac{1}{\tau}\lambda - \mu_k = 0, \quad \text{i.e. } \lambda_k^{\pm} = \frac{1}{2\tau}(-1 \pm \sqrt{1 + 4\mu_k\tau}).$$

Let Λ denote the set of eingenvalues of $\mathbf{A}$. Spontaneous emergence of instabilities is to be expected as soon as the set

$$\sqrt{1 + 4\tau\Lambda} = \left\{ z \in \mathbb{C}, \ \frac{z^2 - 1}{4\tau} \in \Lambda \right\}$$

intersects the half-plane $[\text{Re}(z) > 1]$. An explicit formulation of this condition is proposed in the case of the inertial FTL model, for which the eigenvalues of the first-order model are located on a circle in the half-plane $\text{Re}(z) \leq 0$, tangent at 0 to the imaginary axis. A quite general condition can be proposed, by noticing that the condition above is equivalent to requiring that the set $\tau\Lambda$ intersects the image of $[\text{Re}(z) > 1]$ by the conformal map

$$T : z \longmapsto (z^2 - 1)/4.$$

In other words, we shall preserve stability as far as $\tau\Lambda$ is included in the zone delimited by the parabola $x + y^2 = 0$, that is the image of $[\text{Re}(z) = 1]$ by T. This stability zone is represented in Fig. 10.7 (shaded domain). Note that, if we assume that the first-order model is stable (i.e. we assume that $\Lambda \subset [\text{Re}(z) < 0]$), then all eigenvalues of the inertial model shall be located on the left side of the hyperbola $[x^2 - y^2 = 1]$. Figure 10.7 also represents the FTL case: the circle on the left-hand side represents the dimensionless spectrum $\tau\Lambda$ of the first-order model, and the 8-shaped curve corresponds to the spectrum of the inertial (or delayed) model (up to a τ factor), where

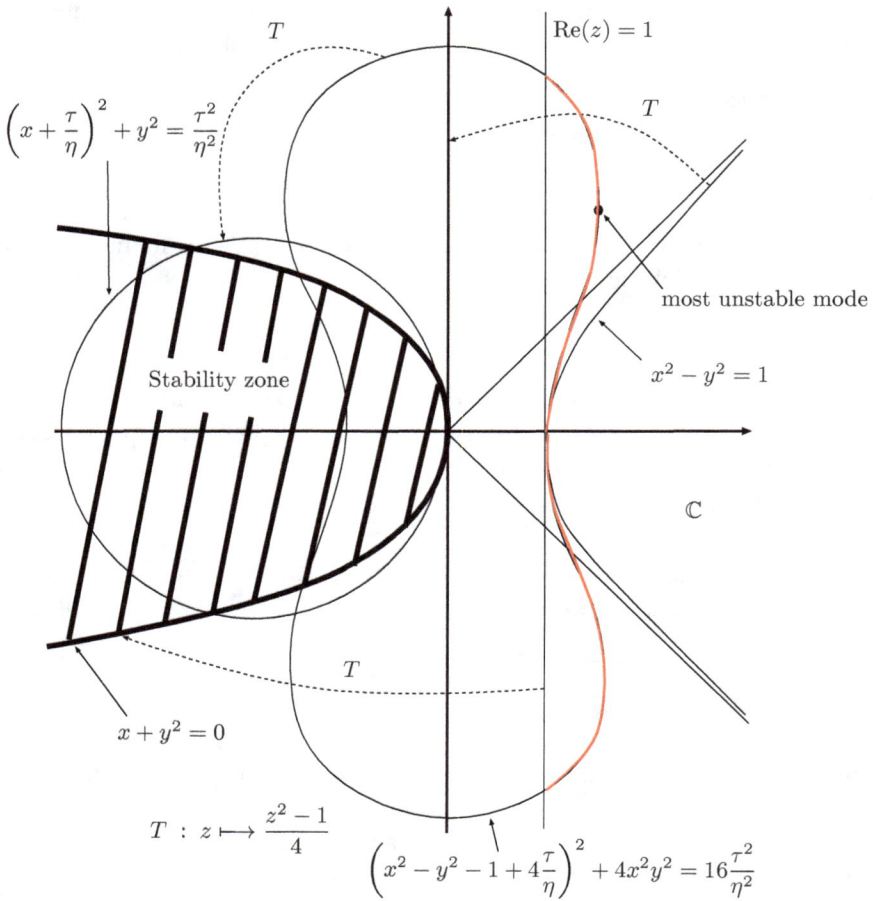

Fig. 10.7.   Stability of second-order systems.

$\eta = 1/\varphi'(w_{\mathrm{eq}})$ is the characteristic time associated to the behavior curve). This curve meets the half-plane $\mathrm{Re}(z) \geq 1$, which ensures that some modes are unstable (red zone). The most unstable mode (i.e. with the largest real part) is indicated as a black dot.

Modeling aspects

Let us interpret this condition in terms of modeling. Consider an eigenvalue $z = x + iy$ of $\mathbf{A}$, associated with a stable mode ($x \leq 0$). In this context it is natural to consider that z (as well as x and y) is expressed in s^{-1} unit (inverse of a time). The imaginary part corresponds (up to a factor $1/2\pi$) to the frequency of oscillations, whereas the modulus of the real part is the inverse of the characteristic relaxation time. The stability condition $\tau y^2 \leq |x|$ can therefore be interpreted as follows: the period of oscillation $T = 2\pi/y$ of the eigenmode has to be larger that 2π times the geometric mean between the relaxation time and the "inertial" parameter τ. Coming back to active entities like pedestrians (or car drivers), we recover the quite intuitive fact that increasing τ, that is increasing the inertial effect or the reaction time of agents (see model (2.19)–(2.20) and Remark 2.13), tends to favor emergence of instabilities.

Appendix

A.1. Ordinary Differential Equations

Existence and uniqueness

Definition A.1 (Cauchy problem). Let $\mathcal{U}$ be an open set of $\mathbb{R}^d$, and $f : \mathcal{U} \times [0, +\infty) \to \mathbb{R}^d$. For any $x_0 \in \mathcal{U}$, the Cauchy problem associated to f and the initial value x_0 consist in finding $t \mapsto x(t) \in U \subset \mathbb{R}^d$ defined over some interval $[0, T)$, such that

$$\begin{cases} \dot{x}(t) = f(x, t), \\ x(t_0) = x_0. \end{cases} \tag{A.1}$$

Definition A.2 (Maximal solution). We call *maximal solution* of the Cauchy problem (A.1) a solution $t \mapsto x(t) \in \mathcal{U}$ defined over an interval $[0, T)$, such that it cannot be extended on a larger interval. More precisely, if $t \mapsto y(t) \in \mathcal{U}$ solves (A.1) over $[0, T')$, and identifies with $x(\cdot)$ over $[0, \min(T, T'))$ then necessarily $T' \leq T$.

Theorem A.1 (Cauchy–Lipschitz). *Under the notation of Definition* A.1, *we assume that f is continuous in $\mathcal{U} \times [0, T)$, and locally Lipschitz continuous with respect to the space variable.[1] Then Cauchy problem (A.1) admits a unique maximal solution defined over $[0, T')$, with $T' \leq T$.*

---

[1] For each $(y, t) \in \mathcal{U} \times [0, T[$, there exist $r > 0$, $\eta > 0$ and $L > 0$ such that

$$|f(y_2, s) - f(y_1, s)| \leq L |y_2 - y_1|$$

for all y_1, y_2 at a distance from y less than r, and for all $s \in [t - \eta, t + \eta]$.

Local/global solutions

Proposition A.1. *In the framework of Theorem A.1, we denote by x the maximal solution, defined over $]0, T^\star[$. If $T^\star < T$ (in other words: if the solution cannot be defined globally), then $x(t)$ necessarily exits all compact sets of U when t goes to $T^\star$, i.e.*
If $T < T^\star$, then

$$\forall K \text{ compact } \subset \mathcal{U}, \ \exists t, \ x(t) \notin K.$$

Proposition A.2. *Suppose that $f : \mathbb{R}^d \times [0, T) \longrightarrow \mathbb{R}^d$ verify the conditions of Theorem A.1, and suppose that there exist A and B such that*

$$|f(x, t)| \leq A |x| + B \quad \text{in } \mathbb{R}^N \times [0, T).$$

Then any maximal solution is global.

Local stability

Definition A.3 (Stability, asymptotic stability). Let x_{eq} be an equilibrium point of the ODE $\dot{x} = f(x)$ in $\mathbb{R}^d$, i.e. x_{eq} such that $f(x_{\text{eq}}) = 0$. We assume that the assumptions of Cauchy–Lipschitz theorem are met in a ball of radius r around x_{eq}. The equilibrium point x_{eq} is said to be

(i) *stable* if for any $0 < \varepsilon < r$, there exists $\eta > 0$ such that, for any x_0 with $|x_0 - x_{\text{eq}}| < \eta$, the solution $t \mapsto x(t)$ associated to the initial solution x_0 remains at a distance from $x(t)$ less than ε;

(ii) *asymptotically stable* if (i) is met, and $|x(t) - x_{\text{eq}}|$ goes to 0 when t goes to $+\infty$.

Theorem A.2. *Let x_{eq} be an equilibrium point of the ODE $\dot{x} = f(x)$ in $\mathbb{R}^d$. We assume that f is continuously differentiable in the neighborhood of x_{eq}. We define*

$$\nabla f = \left(\frac{\partial f_i}{\partial x_j} \right)_{1 \leq i,j \leq N}.$$

(1) *If all the eigenvalues of ∇f have a negative real part, then x_{eq} is asymptotically stable.*

(2) *If one of the eigenvalues of ∇f has a positive real part, then x_{eq} is unstable.*

We refer to Hirsch *et al.* (2017) for a more detailed account of local behavior of solutions to ODE's in the neighborhood of equilibrium points.

Nonlocal stability

Definition A.4 (Lyapunov function). We consider an equilibrium point x_{eq} of the ODE $\dot{x} = f(x)$ in an open set $U \subset \mathbb{R}^N$, where f is assumed to be locally Lipschitz. Let $V \subset U$ be a neighborhood of x_{eq}, that is invariant under the flow of f. We call Lyapunov function for x_{eq} a functional $\Phi : V \mapsto \mathbb{R}$, continuous in a neighborhood $V \subset U$ of x_{eq}, such that

(i) x_{eq} is a strict minimizer of Φ over V,
(ii) $\nabla\Phi(x) \cdot f(x) \leq 0$ for all $x \in V$.

The function Φ is said to be a strict Lyapunov function if, for any $x^0 \in V$, with $x^0 \neq x_{eq}$, the associated solution[2] $t \mapsto x(t)$, with $x(0) = x^0$ such that $\Phi(x(t))$ is strictly decreasing.[3]

Proposition A.3. *With the notation of Definition A.4, if x_{eq} admits a Lyapunov function, then it is stable. If it admits a strict Lyapunov function, then x_0 is globally asymptotically stable, i.e. all trajectories starting in V converge to x_{eq}.*

Numerical approximation of ODE's

All computations presented in this book rely on a very simple algorithm, the so-called explicit Euler scheme. Fixing a time step $\tau > 0$, it consists in building approximations of the solution to (A.1) on the time interval $[0, T]$ at times $0, \tau, \ldots, t^k = k\tau$. Denoting by x^k the approximation of $x(k\tau)$, it reads

$$\frac{x^{k+1} - x^k}{\tau} = f(x^k, t^k), \tag{A.2}$$

with $x^0 = x_0$.

More sophisticated tools to solve such ODE's can be carried out to improve the numerical approximation process in terms of accuracy and/or stability, and we refer to Hairer (1993) for detailed considerations on those improvements.

---

[2]This solution is defined over $[0, +\infty[$ thanks to the Lyapunov functional: it is possible to choose V bounded, so that Proposition A.1 rules out the possibility that the solution may blow up in finite time.

[3]This condition is slightly more general than the condition which is sometimes used in this definition, namely $\nabla V \cdot f(x) < 0$ for any $x \in V \setminus \{x_{eq}\}$.

A.2.   Constrained Optimization

We consider here minimization problems under unilateral constraints. When considering a finite number of affine constraints on $\mathbb{R}^n$, we shall use the matrix formulation $Bv \leq z$, where B is an $m \times n$ matrix, each line of which expresses an affine constraint.

Proposition A.4. *Let U be an open subset of $\mathbb{R}^n$, and $J : U \longmapsto \mathbb{R}$ a continuously differentiable functional. Let B be an $m \times n$ matrix, and $z \in \mathbb{R}^m$. Suppose that $u \in U$ minimizes J on $K = \{v \in U, \ Bv \leq z\}$. Then there exists $\lambda \in \mathbb{R}_+^m$ such that*

$$\begin{cases} \nabla J(u) + B^{\star}\lambda = 0, \\ Bu \leq z, \\ \lambda \cdot Bu = 0. \end{cases}$$

Numerical solution

The computations presented in this book (in particular in Chapter 4) are based on a very simple approach to approximate the minimizer of a quadratic functional

$$v \in \mathbb{R}^n \longmapsto J(v) = \frac{1}{2}Av \cdot v - b \cdot v,$$

over the convex set

$$K = \{v \in \mathbb{R}^n, \ Bu \leq z\},$$

where A is an $n \times n$, symmetric positive definite matrix, $B \in \mathcal{M}_{mn}(\mathbb{R})$, and $z \in \mathbb{R}^m$. The so-called *Uzawa algorithm* is an iterative procedure to approach a solution to the saddle-point formulation of the problem:

$$\begin{cases} Au + B^{\star}\lambda = b, \\ Bu \leq z, \\ \lambda \geq 0, \\ \lambda \cdot Bu = 0. \end{cases} \tag{A.3}$$

Starting from an initial guess $\lambda^0 \in \mathbb{R}^m$, we recursively define

$$\begin{cases} Au^{\ell} + B^{\star}\lambda^{\ell} = b, \\ \lambda^{\ell+1} = \Pi_{+}(\lambda^{\ell} + \rho B(u^{\ell} - z)), \end{cases} \tag{A.4}$$

where $\rho > 0$ is a fixed parameter, and Π_+ is the projection on the cone of vectors with non-negative entries:

$$\Pi_+(\lambda) = (\max(\lambda_i, 0)), \quad 1 \leq i \leq m.$$

The sequence (u^k) can be shown to converge to the minimizer u of J over K, as soon as

$$0 < \rho < 2\alpha/\|B\|^2, \tag{A.5}$$

where α is the smallest eigenvalue of A, i.e. the largest number such that $Av \cdot v \geq \alpha |v|^2$ (see e.g. Ciarlet, 1989).

Bibliography

Argall, B., Cheleshkin, E., Greenberg, J.M., Hinde, C. and Lin, P.-J. (2002). A rigorous treatment of a follow-the-leader traffic model with traffic lights present, *SIAM J. Appl. Math.* **63**(1), pp. 149–168.

Arita, C., Appert-Rolland, C. and Cividini, J. (2015). Two dimensional outflows for cellular automata with shuffle updates, *J. Statist. Mech. Theory Exp.* **2015**.

Aw, A., Klar, A., Materne, T. and Rascle, M. (2002). Derivation of continuum traffic flow models from microscopic follow-the-leader models, *SIAM J. App. Math.* **63**(1), pp. 259–278.

Bellomo, N., Bellouquid, A., Gibelli, L. and Outada, N. (2017). *A Quest Towards a Mathematical Theory of Living Systems Series: Modeling and Simulation in Science, Engineering and Technology*, Birkhaüser, Basel.

Benamou, J.D., Carlier, G. and Santambrogio, F. (2017). Variational mean field games, in *Active Particles*, Advances in Theory, Models, Applications, Vol. 1, eds. (2008). Bellomo, N., Degond, P. and Tadmor, E., Springer, pp. 141–171.

Berthelin, F., Degond, P., Delitala M. and Rascle, M. (2008). A model for the formation and evolution of traffic jams, *Arch. Ration. Mech. Anal.* **187**(2), pp. 185–220.

Blue, V.J. and Adler, J.L. (1998). Emergent fundamental pedestrian flows from cellular automata microsimulation, *Transp. Res. Rec.* **1644**, pp. 29–36.

Brezis, H. (1973). *Opérateurs Maximaux Monotones et Semi-groupes de contractions dans les espaces de Hilbert*, North-Holland, Amsterdam.

Carrillo, J.A., Martin, S. and Wolfram, M.-T. (2016). An improved version of the Hughes model for pedestrian flow, *Math. Models Methods Appl. Sci.* **26**(4), pp. 671–697.

Cepolina, E.M. (2009). Phased evacuation: an optimisation model which takes into account the capacity drop phenomenon in pedestrian flows, *Fire Safety J.* **44**, pp. 532–544.

Chraibi, M., Tordeux, A., Schadschneider, A. (2016) A force-based model to reproduce stop-and-go waves in pedestrian dynamics, in *Traffic and Granular Flow '15*, eds. Knoop V. and Daamen, W., Springer, Cham.

Ciarlet, P.G. (1989). *Introduction to Numerical Linear Algebra and Optimisation*, Cambridge Texts in Applied Mathematics, Cambridge University Press.

Colombo, R.M. and Rossi, E. (2014). On the micro-macro limit in traffic flow, *Rend. Semin. Mat. Univ. Padova* **131**, pp. 217–236.

Corbetta, A., Muntean, A. and Vafayi, K. (2015). Parameter estimation of social forces in pedestrian dynamics models via a probabilistic method, *Math. Biosci. Eng.* **12**(2), pp. 337–356.

Cristiani, E. and Sahu, S. (2016). On the micro-to-macro limit for first-order traffic flow models on networks, *Netw. Heterog. Media* **11**(3), pp. 395–413.

Daamen, W. and Hoogendoorn, S.P. (2012). Emergency door capacity: influence of door width, population composition and stress level, *Fire Technol.* **48**(1), pp. 55–71.

Degond, P., Appert-Rolland, C., Pettre, J. and Theraulaz, G. (2013). Vision-based macroscopic pedestrian models, *Kinet. Relat. Mod.* **6**, pp. 809–839.

Degond, P., Minakowski, P., Navoret, L. and Zatorska, E. (2017). Finite volume approximations of the Euler system with variable congestion, *Comput. Fluids*, https://doi.org/10.1016/j.compfluid.2017.09.007.

Dolak, Y. and Schmeiser, C. (2015). The Keller–Segel model with logistic sensitivity function and small diffusivity, *SIAM J. Appl. Math.* **66**(1), pp. 286–308.

Duives, D.C., Daamen, W. and Hoogendoorn, S.P. (2013). State-of-the-art crowd motion simulation models, *Transp. Res. Part C: Emerging Technol.* **37**, pp. 193–209.

Escobar, R. and Rosa, A. (2003). Architectural design for the survival optimization of panicking fleeing victims, in *ECAL*, Lecture Notes in Computer Science, Vol. 2801, Springer, pp. 97–106.

Faure, S. and Maury, B. (2015). Crowd motion from the granular standpoint, *Math. Models Methods Appl. Sci.* **25**(3), pp. 463–493.

Fiorini, P. and Shiller, Z. (1998). Motion planning in dynamic environments using velocity obstacles, *Int. J. Robotics Res.* **17**(7), pp. 760–772.

Frank, G.A. and Dorso, C.O. (2011). Room evacuation in the presence of an obstacle, *Physica A* **390**, 2135.

Hairer, E. (1993). *Solving Ordinary Differential Equations I. Nonstiff Problems*, Springer Series in Computational Mathematics, Vol. 8, Springer, Berlin.

Garcimartn, A., Zuriguel, I., Pastor, J.M., Martín-Gómez, C. and Parisic, D.R. (2014). Experimental evidence of the "faster is slower" effect, *Transp. Res. Procedia* **6**, pp. 760–767.

Goatin, P. and Scialanga, S. (2016). Well-posedness and finite volume approximations of the LWR traffic flow model with non-local velocity, *Netw. Heter. Media* **11**(1), pp. 107–121.

Goatin, P. and Rossi, F. (2017). A traffic flow model with non-smooth metric interaction: well-posedness and micro-macro limit, *Comm. Math. Sci.* **15**(1), pp. 261–287.

Godlewski, E. and Raviart, P.-A. (1996). Numerical Approximation of Hyperbolic Systems of Conservation Laws, Applied Mathematical Sciences, Vol. 118, Springer, New York.

Hall, E.T. (1969). *The Hidden Dimension*, Anchor Books Editions.

Hartmann, D. and Hasel, P. (2014). Efficient dynamic floor field methods for microscopic pedestrian crowd simulations, *Commun. Comput. Phys.* **16**(1), pp. 264–286.

Helbing, D., Buzna, L., Johansson, A. and Werner, T. (2005). Self-organized pedestrian crowd dynamics: experiments, simulations, and design solutions, *Transp. Sci.* **39**(1), pp. 1–24.

Helbing, D., Farkas I. and Vicsek T. (2000). Simulating dynamical features of escape panic, *Nature* **407**, pp. 487–490.

Helbing, D. and Johansson, A. (2009). Pedestrian, Crowd and evacuation dynamics, in *Encyclopedia of Complexity and Systems Science*, Springer, New York, pp. 6476–6495.

Helbing, D. and Molnár, P. (1995). Social force model for pedestrian dynamics, *Phys. Rev E* **51**, pp. 4282–4286.

Hirsch, M.W., Smale, S. and Devaney, R.L. (2017). *Differential Equations, Dynamical Systems, and an Introduction to Chaos*, third edition, Academic Press.

Hughes R.L. (2002). A continuum theory for the flow of pedestrians, *Transp. Res. Part B* **36**, pp. 507–535.

Hooker, W.W. (1969). On the expected lengths of sequences generated in sorting by replacement selecting, *Commun. ACM*, **12**(7), pp. 411–413.

Hyatt, R.E., Rodarte, J.R., Wilson, T. A. and Lambert, R.K. (1981). Mechanisms of expiratory flow limitation, *Ann. Biomed. Eng.* **9**(5–6), 489–499.

International Maritime Organization (2007). *Guidelines for Evacuation Analysis for New and Existing Passenger Ships*, MSC.1/Circ.1238.

Jelić, A., Appert-Rolland, C., Lemercier, S. and Pettré, J. (2012). Properties of pedestrians walking in line: fundamental diagrams, *Phys. Rev. E* **85**, 036111.

Jiang, L., Li, J., Shen, C., Yang, S. and Han, Z. (2014). Obstacle optimization for panic flow — reducing the tangential momentum increases the escape speed, *PLoS One* **9**(12), e115463.

Keller, E.F. and Segel, L.A. (1970). Initiation of slime mold aggregation viewed as instability, *J. Theor. Biol.* **26**, pp. 399–415.

Keller, E.F. and Segel, L.A. (1971). Traveling bands of chemotactic bacteria: a theoretical analysis, *J. Theor. Biol.* **30**, pp. 235–248.

Kirchner, A., Klüpfel, H., Nishinari, K., Schadschneider, A. and Schreckenberg, M. (2002). Simulation of competitive egress behaviour, *Physica A* **324**, pp. 689–697.

Kirchner, A. and Schadschneider, A. (2002). Simulation of evacuation processes using a bionics-inspired cellular automaton model for pedestrian dynamics, *Physica A* **312**, pp. 260–276.

Kitazawa, K. and Fujiyama, T. (2008). Pedestrian vision and collision avoidance behavior: investigation of the information process space of pedestrians using an eye tracker, in *Pedestrian and Evacuation Dynamics*, eds. Klingsch, W.W.F., Rogsch, C., Schadschneider, A. and Schreckenberg, M., Springer, London, pp. 95–108.

Klüpfel, H., Schreckenberg, M. and Meyer-König, T. (2003). Models for crowd movement and egress simulation, in *Traffic and Granular Flow 03*, eds. Hoogendoorn S.P., Luding S., Bovy P.H.L., Schreckenberg, M. and Wolf D.E., Springer, Berlin.

Kretz, T. (2015). On oscillations in the social force model, *Physica A* **438**, pp. 272–285.

Kretz, T. and Schreckenberg, M. (2006). Moore and more and symmetry, in *Pedestrian and Evacuation Dynamics 2005*, eds. Waldau *et al.*, Springer, Berlin, pp. 297–308.

Lasry, J.M. and Lions, P.L. (2007). Mean field games, *Japanese J. Math.* **2**(1), 229–260.

Lemercier, S., Jelić, A., Kulpa, R., Hua, J., Fehrenbach, J., Degond, P., Appert-Rolland, C., Donikian, C. and Pettr, J. (2012). Realistic following behaviors for crowd simulation, *Comput. Graphics Forum* **31**(2), pp. 489–498.

Levin, B.M. (1984). Human behavior in fire: what we know now, in *SFPE Fire Protection Engineering Seminars*, New Orleans, May 21–24, SFPE Technology Report 84-3.

Lighthill, M.J. and Whitham, G.B. (1955). On kinematic waves II: a theory of traffic flow on long, crowded roads, *Proc. Roy. Soc. London Ser. A* **229**, pp. 317–345.

Lin, P., Ma, J., Liu, T.Y., Ran, T., Si, Y.L. , Wu, F.Y. and Wang, G.Y. (2017). An experimental study of the impact of an obstacle on the escape efficiency by using mice under high competition, *Physica A* **482**, pp. 228–242.

Maury, B. (2006). A time-stepping scheme for inelastic collisions, *Numer. Math.* **102**(4), pp. 649–679.

Maury, B. (2016). Congested transport at microscopic and macroscopic scales, in *7th European Congress of Mathematics (7ECM)*, Berlin.

Maury, B., Roudneff-Chupin, A. and Santambrogio, F. (2010). A macroscopic crowd motion model of gradient flow type, *Math. Models Methods Appl. Sci.* **20**, pp. 1787–1821.

Maury, B., Roudneff-Chupin, A., Santambrogio, F. and Venel, J. (2011). Handling congestion in crowd motion modeling, *Netw. Heterog. Media* **6**(3), pp. 485–519.

Maury, B. and Venel, J. (2011). A discrete contact model for crowd motion, *ESAIM: M2AN* **45**(1).

Mirebeau, J.M. (2017). Fast Marching methods for curvature penalized shortest paths, *J. Math. Imaging Vis.* https://doi.org/10.1007/s10851-017-0778-5.

Mirebeau, J.M., Hamiltonian fast Marching, Software, Github repository, https://github.com/Mirebeau.

Moreau, J.-J. (1962). Décomposition orthogonale d'un espace Hilbertien selon deux cônes mutuellement polaires, *C. R. Acad. Sci. Paris* **255**, pp. 238–240.

Moreau, J.J. (1977). Evolution problem associated with a moving convex set in a *Hilbert* space, *J. Differential Equations* **26**(3), pp. 347–374.

Nicolas, A., Bouzat, S. and Kuperman, M.N. (2016). Statistical fluctuations in pedestrian evacuation times and the effect of social contagion, *Phys. Rev. E* **94**.

Pastor, J.M., Garcimartín, A., Gago, P. A., Peralta, J. P., Martín-Gómez, C., Ferrer, L.M., Maza, D., Parisi, D.R., Pugnaloni, L.A. and Zuriguel, I. (2015). Experimental proof of faster-is-slower in systems of frictional particles flowing through constrictions *Phys. Rev. E* **92**.

Portz, A. and Seyfried, A. (2010). Modeling stop-and-go waves in pedestrian dynamics, in *International Conference on Parallel Processing and Applied Mathematics, PPAM 2009: Parallel Processing and Applied Mathematics*, Springer, Berlin, pp. 561–568.

Portz, A. and Seyfried, A. (2011). Analyzing stop-and-go waves by experiment and modeling, in *Pedestrian and Evacuation Dynamics*, Springer US, pp. 577–586.

Rastogi, R., Ilango, T. and Chandra, S. (2013). Pedestrian flow characteristics for different pedestrian facilities and situations, *European Transport Trasporti Europei*, April 2013, Issue 53, Paper No 6.

Richards, P.I. (1956). Shock waves on the highway, *Oper. Res.* **4**, pp. 42–51.

Rogsch, C., Klingsch, W., Seyfried, A. and Weigel, H. (2009). Prediction accuracy of evacuation times for high-rise buildings and simple geometries by using different software-tools, in *Traffic and Granular Flow 2007*, pp. 395–400.

Rouphail, N.M. and Allen, D.P. (1998). Capacity analysis of pedestrian and bicycle facilities, recommended procedures chapter 13, pedestrians, of the highway capacity manual, Report No. FHWA-RD-98-108.

Santambrogio, F. (2015). *Optimal Transport for Applied Mathematicians*, Birkhäuser.

Schadschneider, A. (2001). Cellular automaton approach to pedestrian dynamics — theory, in *Pedestrian and Evacuation Dynamics*, eds. Schreckenberg, M. and Sharma S.D., Springer, p. 75.

Schadschneider, A. and Seyfried, A. (2011). Empirical results for pedestrian dynamics and their implications for modeling, *Netw. Heterog. Media* **6**, pp. 545–560.

Schadschneider, A. and Seyfried, A. (2011). Empirical results for pedestrian dynamics and their implications for cellular automata models, in *Pedestrian Behavior*, ed. Timmermans, H., Emerald, p. 27.

Sethian, J.A. (1999). *Level Set Methods and Fast Marching Methods: Evolving Interfaces in Computational Geometry, Fluid Mechanics, Computer Vision, and Materials Science*, Cambridge University Press.

Seyfried, A., Steffen, B., Klingsch, W. and Boltes, M. (2005). The fundamental diagram of pedestrian movement revisited, *J. Statist. Mech. Theory Exp.* **2005**.

Starke, J., Thomsen, K.B., Sørensen, A., Marschler, C., Schilder, F., Dederichs, A. and Hjorth, P. (2014). Nonlinear effects in examples of crowd evacuation scenarios, in *2014 IEEE 17th International Conference on Intelligent Transportation Systems (ITSC)*, October 8–11, Qingdao, China.

Tantucci, C., Duguet, A., Giampiccolo, P., Similowski, T., Zelter, M. and Derenne, J.P. (2002). The best peak expiratory flow is flow-limited and effort-independent in normal subjects, *Am. J. Respir. Crit. Care Med.* **165**(9), pp. 1304–1308.

Thibault, L. (2003). Sweeping process with regular and nonregular sets, *J. Differential Equations* **193**, pp. 1–26.

Twarogowska, M., Goatin, P. and Duvigneau, R. (2014). Comparative study of macroscopic pedestrian models, *Transp. Res. Procedia* **2**, pp. 477–485. *The Conference on Pedestrian and Evacuation Dynamics 2014 (PED2014)*.

Van den Berg, J., Lin, M. and Manocha, D. (2008). Reciprocal Velocity Obstacles for real-time multi-agent navigation, *IEEE Int. Conf. Robotics and Automation*, pp. 1928–1935.

Van den Berg, J., Guy, S., Lin, M. and Manocha, D. (2011). Reciprocal n-body collision avoidance, *Robotics Res.* **70**, pp. 3–19.

Venel, J. (2011). A numerical scheme for a class of sweeping processes, *Numer. Math.* **118**(2), pp. 367–400.

Villani, C. (2003). *Topics in Optimal Transportation*, Graduate Studies in Mathematics, Vol. 58, American Mathematical Society, Providence, RI.

Wang, P. (2016). Understanding social-force model in psychological principles of collective behavior, arXiv: e 1605.05146.

Weidmann, U. (1992). Transporttechnik der Fußgänger, in *Schriftenreihe des IVT 90*, ETH Zürich (in German).

Williams, J.W.J. (1964). Algorithm 232 — Heapsort, *Commun. ACM* **7**(6), pp. 347–348.

Xiaoping, Z., Wei, L. and Chao, G. (2010). Simulation of evacuation processes in a square with a partition wall using a cellular automaton model for pedestrian dynamics, *Physica A* **389**, pp. 2177–2188.

Yanagisawa, D., Kimura, A., Tomoeda, A., Nishi, R., Suma, Y., Ohtsuka, K. and Nishinari, K. (2009). Introduction of frictional and turning function for pedestrian outflow with an obstacle, *Phys. Rev. E* **80**(3 Pt 2), Art. No. 036110.

Zuriguel, I., Olivares, J., Pastor, J.M., Martín-Gómez, C., Ferrer, L.M., Ramos, J.J. and Garcimartín, A. (2016). Effect of obstacle position in the flow of sheep through a narrow door, *Phys. Rev. E* **94**.

Index

www.ingramcontent.com/pod-product-compliance
Lightning Source LLC
Chambersburg PA
CBHW050627190326
41458CB00008B/2167